Optimization for Industrial Problems

Patrick Bangert

Optimization for Industrial Problems

 Springer

Patrick Bangert
algorithmica technologies GmbH
Bremen, Germany

ISBN 978-3-642-24973-0   e-ISBN 978-3-642-24974-7
DOI 10.1007/978-3-642-24974-7
Springer Heidelberg Dordrecht London New York

Library of Congress Control Number: 2011945031

Mathematics Subject Classification (2010): 90-08, 90B50

Springer is part of Springer Science+Business Media (www.springer.com)

It can be done !

algorithmica technologies GmbH
Advanced International Research Institute on Industrial
Optimization gGmbH
Department of Mathematics, University College London

Preface

Some Early Opinions on Technology

There is practically no chance communications space satellites will be used to provide better telephone, telegraph, television, or radio service inside the United States

T. Craven, FCC Commissioner, 1961

There is not the slightest indication that nuclear energy will ever be obtainable. It would mean that the atom would have to be shattered at will.

Albert Einstein, 1932

Heavier-than-air flying machines are impossible.

Lord Kelvin, 1895

We will never make a 32 bit operating system.

Bill Gates, 1983

Such startling announcements as these should be deprecated as being unworthy of science and mischievous to its true progress.

William Siemens, on Edison's light bulb, 1880

The energy produced by the breaking down of the atom is a very poor kind of thing. Anyone who expects a source of power from the transformation of these atoms is talking moonshine.

Ernest Rutherford, shortly after splitting the atom for the first time, 1917

Everything that can be invented has been invented.

Charles H. Duell, Commissioner of the US Patent Office, 1899

Content and Scope

Optimization is the determination of the values of the independent variables in a function such that the dependent variable attains a maximum over a suitably defined area of validity (c.f. the boundary conditions). We consider the case in which the independent variables are many but the dependent variable is limited to one; multi-criterion decision making will only be touched upon.

This book, for the first time, combines mathematical methods and a wide range of real-life case studies of industrial use of these methods. Both the methods and the problems to which they are applied as examples and case studies are useful in real situations that occur in profit making industrial businesses from fields such as chemistry, power generation, oil exploration and refining, manufacturing, retail and others.

The case studies focus on real projects that actually happened and that resulted in positive business for the industrial corporation. They are problems that other companies also have and thus have a degree of generality. The thrust is on take-home lessons that industry managers can use to improve their production via optimization methods.

Industrial production is characterized by very large investments in technical facilities and regular returns over decades. Improving yield or similar characteristics in a production facility is a major goal of the owners in order to leverage their investment. The current approach to do this is mostly via engineering solutions that are costly, time consuming and approximate.

Mathematics has entered the industrial stage in the 1980s with methods such as linear programming to revolutionize the area of industrial optimization. Neural networks, simulation and direct modeling joined and an arsenal of methods now exists to help engineers improve plants; both existing and new. The dot-com revolution in the late 1990s slowed this trend of knowledge transfer and it is safe to say that the industry is essentially stuck with these early methods. Mathematics has evolved since then and accumulated much expertise in optimization that remains hardly used. Also, modern computing power has exploded with the affordable parallel computer so that methods that were once doomed to the dusty shelf can now actually be used.

These two effects combine to harbor a possible revolution in industrial uses for mathematical methods. These uses center around the problem of optimization as almost every industrial problem concerns maximizing some goal function (usually efficiency or yield). We want to help start this revolution by a coordinated presentation of methods, uses and successful examples.

The methods are necessarily heuristic, i.e. non-exact, as industrial problems are typically very large and complex indeed. Also, industrial problems are defined by imprecise, sometimes even faulty data that must be absorbed by a model. They are always non-linear and have many independent variables. So we must focus on heuristic methods that have these characteristics.

This book is practical

This book is intended to be used to solve real problems in a handbook manner. It should be used to look for potential yet untapped. It should be used to see possibilities where there were none before. The impossible should move towards the realm of the possible. The use, therefore, will mainly be in the sphere of application by persons employed in the industry.

The book may also be used as instructional material in courses on either optimization methods or applied mathematics. It may also be used as instructional material in MBA courses for industrial managers.

Many readers will get their first introduction as to what mathematics can really and practically do for the industry instead of general commonplaces. Many will find out what problems exist where they previously thought none existed. Many will discover that presumed impossibilities have been solved elsewhere. In total, I believe that you, the reader, will benefit by being empowered to solve real problems.

These solutions will save the corporations money, they will employ people, they will reduce pollution into the environment. They will have impact. It will show people also that very theoretical sciences have real uses.

It should be emphasized that this book focuses on applications. Practical problems must be understood at a reasonable level before a solution is possible. Also all applications have several non-technical aspects such as legal, compliance and managerial ramifications in addition to the obvious financial dimension. Every solution must be implemented by people and the interactions with them is the principal cause for failure in industrial applications. The right change management including the motivation of all concerned is an essential element that will also be addressed. Thus, this book presents cases as they can really function in real life.

Due to the wide scope of the book, it is impossible to present neither the methods nor the cases in full detail. We present what is necessary for understanding. To actually implement these methods, a more detailed study or prior knowledge is required. Many take-home lessons are however spelt out. The major aim of the book is to generate understanding and not technical facility.

This book is intended for practitioners

The intended readership has five groups:

1. *Industrial managers* - will learn what can be done with mathematical methods. They will find that a lot of their problems, many seemingly impossible, are already solved. These methods can then be handed to technical persons for implementation.
2. *Industrial scientists* - will use the book as a manual for their jobs. They will find methods that can be applied practically and have solve similar problems before.
3. *University students* - will learn that their theoretical subjects do have practical application in the context of diverse industries and will motivate them in their studies towards a practical end. As such it will also provide starting points for theses.
4. *University researchers* - will learn to what applications the methods that they research about have been put or respectively what methods have been used by others to solve problems they are investigating. As this is a trans-disciplinary

book, it should facilitate communication across the boundaries of the mathematics, computer science and engineering departments.

5. *Government funding bodies* - will learn that fundamental research does actually pay off in many particular cases.

A potential reader from these groups will be assumed to have completed a mathematics background training up to and including calculus (European high-school or US first year college level). All other mathematics will be covered as far as needed. The book contains no proofs or other technical material; it is practical.

A short summary

Before a problem can be solved, it and the tools must be understood. In fact, a correct, complete, detailed and clear description of the problem is (measured in total human effort) often times nearly half of the final solution. Thus, we will place substantial room in this book on understanding both the problems and the tools that are presented to solve them.

Indeed we place primary emphasis on understanding and only secondary emphasis on use. For the most part, ready-made packages exist to actually perform an analysis. For the remainder, experts exist that can carry it out. What cannot be denied however, is that a good amount of understanding must permeate the relationship between the problem-owner and the problem-solver; a relationship that often encompasses dozens of people for years.

Here is a brief list of the contents of the chapters

1. What is optimization?
2. What is an optimization problem?
3. What are the management challenges in an optimization project?
4. How can we deal with faulty and noisy empirical data?
5. How do we gain an understanding of our dataset?
6. How is a dataset converted into a mathematical model?
7. How is the optimization problem actually solved?
8. What are some challenges in implementing the optimal solution in industrial practice (change management)?

Most of the book was written by me. Any deficiencies are the result of my own limited mind and I ask for your patience with these. Any benefits are, of course, obtained by standing on the shoulders of giants and making small changes. Many case studies are co-authored by the management from the relevant industrial corporations. I heartily thank all co-authors for their participation! All the case studies were also written by me and the same comments apply to them. I also thank the co-authors very much for the trust and willingness to conduct the projects in the first place and also to publish them here.

Chapter 8 was entirely written by Andreas Ruff of Elkem Silicon Materials. He has many years of experience in implementing optimization projects' results in chemical corporations and has written a great practical account of the potential pitfalls and their solutions in change management.

Following this text, we provide first an alphabetical list of all co-authors and their affiliations and then a list of all case studies together with their main topics and educational illustrations.

Bremen, 2011   *Patrick Bangert*

Markus Ahorner
COO
m.ahorner@algorithmica-technologies.com
Section 4.8, p. 53; Section 4.9, p. 58

algorithmica technologies GmbH
Gustav-Heinemann-Strasse 101
28215 Bremen, Germany
www.algorithmica-technologies.com

Dr. Patrick Bangert
CEO
p.bangert@algorithmica-technologies.com
All sections

algorithmica technologies GmbH
Gustav-Heinemann-Strasse 101
28215 Bremen, Germany
www.algorithmica-technologies.com

Claus Borgböhmer
Director Project Management
claus.borgboehmer@de.sasol.com
Section 4.8, p. 53

Sasol Solvents Germany GmbH
Römerstrasse 733
47443 Moers, Germany
www.sasol.com

Pablo Cajaraville
Director Engineering and Sales
p.cajaraville@reinermicrotek.com
Section 6.6, p. 135

Reiner Microtek
Poligono Industrial Itziar, Parcela H-3
20820 Itziar-Deba, Spain
www.reinermicrotek.com

Roger Chevalier
Senior Research Engineer
roger.chevalier@edf.fr
Section 6.10, p. 152

EDF SA, R&D Division
6 quai Watier, BP49
78401 Chatou Cedex, France
www.edf.com

Jörg-A. Czernitzky
Power Plant Group Director Berlin
joerg-andreas.czernitzky@vattenfall.de
Section 7.10, p. 197

Vattenfall Europe Wärme AG
Puschkinallee 52
12435 Berlin, Germany
www.vattenfall.de

Prof. Dr. Adele Diederich
Professor of Psychology
a.diederich@jacobs-university.de
Section 7.6, p. 183

Jacobs University Bremen gGmbH
P.O. Box 750 561
28725 Bremen, Germany
www.jacobs-university.de

Björn Dormann
Research Director
b.dormann@desma.de
Section 6.6, p. 135

Klöckner Desma Schuhmaschinen GmbH
Desmastr. 3/5
28832 Achim, Germany
www.desma.de

Hans Dreischmeier
Director SAP
hans.dreischmeier@vestolit.de
Section 4.9, p. 58

Vestolit GmbH & Co. KG
Industriestrasse 3
45753 Marl, Germany
www.vestolit.de

Bernd Herzog
Quality Control Manager
bernd.herzog@hella.de
Section 4.11, p. 63

Hella Fahrzeugkomponenten GmbH
Dortmunder Strasse 5
28199 Bremen, Germany
www.hella.de

Dr. Philipp Imgrund
Director Biomaterial Technology
Director Power Technologies
philipp.imgrund@ifam.fraunhofer.de
Section 6.6, p. 135

Fraunhofer Institute for Manufacturing
and Advanced Materials IFAM
Wiener Strasse 12
28359 Bremen, Germany
www.ifam.fhg.de

Maik Köhler
Technical Expert
m.koehler@desma.de
Section 6.6, p. 135

Klöckner Desma Schuhmaschinen GmbH
Desmastr. 3/5
28832 Achim, Germany
www.desma.de

Lutz Kramer
Project Manager
Metal Injection Molding
lutz.kramer@ifam.fraunhofer.de
Section 6.6, p. 135

Fraunhofer Institute for Manufacturing
and Advanced Materials IFAM
Wiener Strasse 12
28359 Bremen, Germany
www.ifam.fhg.de

Guisheng Li
Institute Director
lgsh50@tom.com
Section 6.12, p. 157

Oil Production Technology Research Institute
Plant No. 5 of Petrochina Dagang Oilfield Company
Tianjin 300280, China
www.petrochina.com.cn

Bailiang Liu
Vice Director
liubailiang@petrochina.com.cn
Section 7.9, p. 194

PetroChina Dagang Oilfield Company
Tianjin 300280
China
www.petrochina.com.cn

Oscar Lopez   MIM TECH ALFA, S.L.
Senior Research Engineer   Avenida Otaola, 4
olopez@alfalan.es   20600 Eibar, Spain
Section 6.6, p. 135   www.alfalan.es

Torsten Mager   KNG Kraftwerks- und Netzgesellschaft mbH
Director Technical Services   Am Kühlturm 1
torsten.mager@kng.de   18147 Rostock, Germany
Section 5.6, p. 102   www.kraftwerk-rostock.de

Manfred Meise   Hella Fahrzeugkomponenten GmbH
CEO   Dortmunder Strasse 5
manfred.meise@hella.de   28199 Bremen, Germany
Section 4.11, p. 63   www.hella.de

Kurt Müller   Vestolit GmbH & Co. KG
Director Maintenance   Industriestrasse 3
kurt.mueller@vestolit.de   45753 Marl, Germany
Section 4.9, p. 58   www.vestolit.de

Kaline Pagnan Furlan   Fraunhofer Institute for Manufacturing
Research Assistant   and Advanced Materials IFAM
Metal Injection Molding   Wiener Strasse 12
kaline.pagnan.furlan@ifam.fraunhofer.de   28359 Bremen, Germany
Section 6.6, p. 135   www.ifam.fhg.de

Yingjun Qu   Oil Production Technology Research Institute
Institute Director   Plant No. 6 of Petrochina Changqing Oilfield Company
jyj_cq@petrochina.com.cn   718600 Shanxi, China
Section 6.12, p. 157   www.petrochina.com.cn

Pedro Rodriguez   MIM TECH ALFA, S.L.
Director R&D   Avenida Otaola, 4
prodriguez@alfalan.es   20600 Eibar, Spain
Section 6.6, p. 135   www.alfalan.es

Andreas Ruff   Elkem Silicon Materials
Technical Marketing Manager   Hochstadenstrasse 33
andreas.ruff@elkem.no   50674 Kln
Chapter 8, p. 201   www.elkem.no

Dr. Natalie Salk   PolyMIM GmbH
CEO   Am Gefach

natalie.salk@polymim.com   55566 Bad Sobernheim
Section 6.6, p. 135   www.polymim.com

Prof. Chaodong Tan   China University of Petroleum
Professor   Beijing 102249
tantcd@126.com   China
Section 6.12, p. 157; Section 7.9, p. 194   www.upc.edu.cn

Jörg Volkert   Fraunhofer Institute for Manufacturing
Project Manager   and Advanced Materials IFAM
Metal Injection Molding   Wiener Strasse 12
joerg.volkert@ifam.fraunhofer.de   28359 Bremen, Germany
Section 6.6, p. 135   www.ifam.fhg.de

Xuefeng Yan   Beijing Yadan Petroleum Technology Co., Ltd.
Director of Production Technology   No. 37 Changqian Road, Changping
yxf_993@126.com   Beijing 102200, China
Section 6.12, p. 157   www.yadantech.com

Jie Zhang   Yadan Petroleum Technology Co Ltd
Vice CEO   No. 37 Changqian Road, Changping
nyboc@sina.com   Beijing 102200, China
Section 7.9, p. 194   www.yadantech.com

Timo Zitt   RWE Power AG
Director Dormagen Combined-Cycle Plant   Chempark, Geb. A789
timo.zitt@rwe.com   41538 Dormagen
Section 7.11, p. 199   www.rwe.com

The following is a list of all case studies provided in the book. For each study, we provide its location in the text and its title. The summary indicates what the case deals with and what the result was. The "lessons" are the mathematical optimization concepts that this case particularly illustrates.

Self-Benchmarking in Maintenance of a Chemical Plant
Section 4.8, p. 53
Summary: In addition to the common practice of benchmarking, we suggest to compare the plant to itself in the past to make a self-benchmark.
Lessons: The right pre-processing of raw data from the ERP system can already bear useful information without further mathematical analysis.

Financial Data Analysis for Contract Planning
Section 4.9, p. 58

Summary: Based on past financial data, we create a detailed projection into the future in several categories and so provide decision support for budgeting.
Lessons: Discovering basic statistical features of data first, allows the transformation of ERP data into a mathematical framework capable of making reliable projections.

Early Warning System for Importance of Production Alarms
Section 4.11, p. 63
Summary: Production alarms are analyzed in terms of their abnormality. Thus we only react to those alarms that indicate qualitative change in operations.
Lessons: Comparison of statistical distributions based on statistical testing allows us to distinguish normal from abnormal events.

Optical Digit Recognition
Section 5.4, p. 92
Summary: Images of hand-written digits are shown to the computer in an effort for it to learn the difference between them without us providing this information (unsupervised learning).
Lessons: It is possible to cluster data into categories without providing any information at all apart from the raw data but it pays to pre-process this data and to be careful about the number of categories specified.

Turbine Diagnosis in a Power Plant
Section 5.5, p. 96
Summary: Operational data from many turbines are analyzed to determine which turbine was behaving strangely and which was not.
Lessons: Time-series can be statistically compared based on several distinctive features providing an automated check on qualitative behavior of the system.

Determining the Cause of a Known Fault
Section 5.6, p. 102
Summary: We search for the cause of a bent blade of a turbine and do not find it.
Lessons: Sometimes the causal mechanism is beyond current data acquisition and then cannot be analyzed out of it. It is important to recognize that analysis can only elucidate what is already there.

Customer Segmentation
Section 5.10, p. 117
Summary: Consumers are divided into categories based on their purchasing habits.
Lessons: Based on purchasing histories, it is possible to group customers into behavioral groups. It is also possible to extract cause-effect information about which purchases trigger other purchases.

Scrap Detection in Injection Molding Manufacturing
Section 6.6, p. 135

Summary: It is determined whether an injection molded part is scrap or not.
Lessons: Several time-series need to be converted into a few distinctive features to then be categorized by a neural network as scrap or not.

Prediction of Turbine Failure
Section 6.7, p. 140
Summary: A turbine blade tear is correctly predicted two days before it happened.
Lessons: Time-series can be extrapolated into the future and thus failures predicted. The failure mechanism must be visible already in the data.

Failures of Wind Power Plants
Section 6.8, p. 143
Summary: Failures of wind power plants are predicted several days before they happen.
Lessons: Even if the physical system is not stable because of changing wind conditions, the failure mechanism is sufficiently predictable.

Catalytic Reactors in Chemistry and Petrochemistry
Section 6.9, p. 148
Summary: The catalyst deactivation in fluid and solid catalytic reactors is projected into the future.
Lessons: Non-mechanical degradation can be predicted as well and allows for projection over one year in advance.

Predicting Vibration Crises in Nuclear Power Plants
Section 6.10, p. 152
Summary: A temporary increase in turbine vibrations is predicted several days before it happens.
Lessons: Subtle events that are not discrete failures but rather quantitative changes in behavior can be predicted too.

Identifying and Predicting the Failure of Valves
Section 6.11, p. 155
Summary: In a system of valves, we determine which valve is responsible for a non-constant final mixture and predict when this state will be reached.
Lessons: Using data analysis in combination with plant know-how, we can identify the root-cause even if the system is not fully instrumented.

Predicting the Dynamometer Card of a Rod Pump
Section 6.12, p. 157
Summary: The condition of a rod pump can be determined from a diagram known as the dynamometer card. This 2D shape is projected into the future in order to diagnose and predict future failures.

Lessons: It is possible not only to predict time-series but also changing geometrical shapes based on a combination of modeling and prediction.

Human Brains use Simulated Annealing to Think
Section 7.6, p. 183
Summary: Based on human trial, we determine that human problem solving uses the simulated annealing paradigm.
Lessons: Simulated annealing is a very general and successful method to solve optimization problems that, when combined with the natural advantages of the computer, becomes very powerful and can find the optimal solution in nearly all cases.

Optimization of the Müller-Rochow Synthesis of Silanes
Section 7.8, p. 189
Summary: A complex chemical reaction whose kinetics is not fully understood by science is modeled with the aim of increasing both selectivity and yield.
Lessons: It is possible to construct empirical models without theoretical understanding and still compute the desired answers.

Increase of Oil Production Yield in Shallow-Water Offshore Oil Wells
Section 7.9, p. 194
Summary: Offshore oil pumps are modeled with the aim of both predicting their future failures and increasing the oil production yield.
Lessons: The pumps must be considered as a system in which the pumps influence each other. We solve a balancing problem between them using their individual models.

Increase of coal burning efficiency in CHP power plant
Section 7.10, p. 197
Summary: The efficiency of a CHP coal power plant is increased by 1%.
Lessons: While each component in a power plant is already optimized, mathematical modeling offers added value in optimizing the combination of these components into a single system. The combination still allows a substantial efficiency increase based on dynamic reaction to changing external conditions.

Reducing the Internal Power Demand of a Power Plant
Section 7.11, p. 199
Summary: A power plant uses up some its own power by operating pumps and fans. The internal power is reduced by computing when these should be turned off.
Lessons: We extrapolate discrete actions (turning off and on of pumps and fans) from the continuous data from the plant in order to optimize a financial goal.

Contents

Chapter 1
Overview of Heuristic Optimization

1.1 What is Optimization?

Suppose we have a function $f(x)$ where the variable x may be a vector of many dimensions. We seek the point x^* such that $f(x^*)$ is the maximum value among all possible $f(x)$. This point x^* is called the *global optimum* of the function $f(x)$. It is possible that x^* is a unique point but it is also possible that there are several points that share the maximal value $f(x^*)$. *Optimization* is a field of mathematics that concerns itself with finding the point x^* given the function $f(x)$.

There are two fine distinctions to be made relative to this. First, the point x^* is the point with highest $f(x)$ for all possible x and as such the global optimum. We are usually interested in this global optimum. There exists the concept of a *local optimum* that is the point with highest $f(x)$ for all x in the neighborhood of the local optimum. For example, any peak is a local optimum but only the highest peak is the global maximum. Usually we are not interested in finding local optima but we are interested in recognizing them because we want to be able to determine that, while we are on a peak, there exists a higher peak elsewhere.

Second, the phrase "all possible x needs careful consideration. Usually any value of the independent variable is allowed ", $x \in [-\infty, \infty]$, but in some cases the independent variable is restricted. Such restrictions may be very simple like $3 \leq x \leq 18$. Some may be complex by not giving explicit limitations but rather tying two elements of the independent variable vector together, e.g.

$$x_1 < \int_0^\infty g(x_2) dx$$

where $g(x)$ is some other function. Any such equation is called a *constraint* or *boundary condition*.

Almost all practical optimization problems are constrained or bounded. Simple boundaries are usually no problem. Complicated constraints like the integral constraint above are usually complex and must often be treated specially.

P. Bangert (ed.), *Optimization for Industrial Problems*,
DOI 10.1007/978-3-642-24974-7_1, © Springer-Verlag Berlin Heidelberg 2012

1.1.1 Searching vs. Optimization

Consider a map of a mountain range. The location variable x is a two-component vector where the two components are latitude and longitude. The function $f(x)$ is the altitude corresponding to the particular location. The task is now to find x^*, i.e. the point (on the map) with the highest altitude.

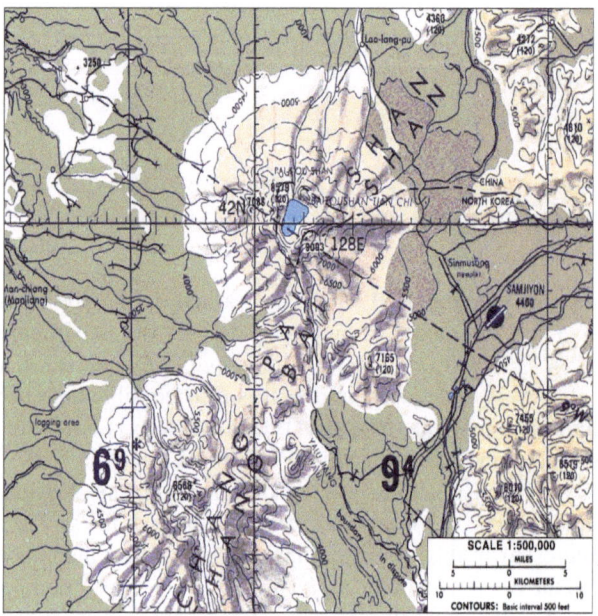

Fig. 1.1 A topographical map of the Baitoushan mountain range in China. The contours are labeled with the altitude in this case.

As humans, we usually accomplish this by searching. If the map is a topological map (see figure 1.1), we would generally use the contour lines to aid our search knowing that the centers of roughly circular contours are bound to be mountains. In the absence of visual aids like contours or colored shading, we have to rely on searching. We know that we can get a reasonable guess by random searching but the only sure way to find the highest peak is by exhaustively reading all the labels on the map.

Moreover, the map does not allow us to find peaks that are not on the map even though they may be even higher. This is a practical example of a boundary condition. We know that the highest peak on Earth is Mount Everest but that, on a map of Europe, we will find Mont Blanc to be the highest peak instead. Because the boundary of Europe excluded Mount Everest, we are not able to find it but we did find the best point satisfying the boundary conditions.

This example illustrates the two principal problems of optimization: (1) the incorporation of boundary conditions or constraints and (2) the inherent search nature of the problem. Practically, we must specify two more items: (3) the numerical accuracy with which x^* must be determined[1] and (4) the lowest probability that we are prepared to accept of the final answer actually being the true global optimum.

In the following sections, we will analyze each of these in turn.

1.1.2 Constraints

As introduced at the beginning of the section, constraints limit the allowed range of the independent variables. In practical situations, we will almost always have an upper and lower limit to any independent variable, e.g. $-1 \leq x_1 \leq 2.5$. Many times such limits indicate the normal operating conditions of a piece of equipment or the safety limits for operation thereof. Such limits are often contained in process control systems or safety systems and are usually determined during the engineering and build phase of a production plant's lifetime.

In addition to these simple constraints that limit the numerical range of each variable, we generally have limitations on the interdependency between variables. These are generally quite application specific and are difficult to discuss in general. An example of such is that the total flow through a system of devices operated in parallel must be (roughly) equal to the sum of the individual flows. Placing limits on any one of these interdependent variables (e.g. requiring a specific total flow) will induce interesting and non-trivial limits on the other variables as there is now some, but not total, flexibility.

It goes without saying that if constraints are not specified, they are generally not met by any computational method. Thus, it is vital to specify these, if they exist or are necessary.

1.1.3 Finding through a little Searching

Briefly, optimization is the art of finding without searching (much). In our efforts to find the best point, we must perform a search of some kind. The most primitive search possible is to examine every location and, at the end, output the best point found. The advantages of this method are that it is simple, always succeeds and is very general. The disadvantage is that it is generally slow as there are many possible points to look through.

We must therefore come up with a way to find what we need without looking through all possibilities. In fact, as the number of possibilities grows rapidly with

---

[1] The location of the Mount Everest's peak must be specified more accurately for a mountain climber than for a photo journalist – demanded accuracy depends on the intended application. More demanded accuracy may lead to a lot more work required to determine it.

the problem size, *we must come up with a way that needs us to examine only very very few of all the possibilities*. This is the first important insight of optimization.

At first sight, we might think that this is a task that requires problem specific knowledge. Indeed, domain knowledge is very useful and should, in general, be used to every extent possible. However, it is possible to say many things in total absence of domain knowledge. It is this that will be the major topic of this book.

1.1.4 Accuracy

Whenever we ask a question whose answer is a number, we must really specify how accurately we need that number. Suppose that an engineer wants to make a calculation on a design and requires the number π. That $\pi \approx 3 \pm 0.2$ is correct to a certain degree of accuracy but is usually not accurate enough for engineering purposes. That $\pi \approx 3.14 \pm 0.002$ is better but may still not be good enough. This can be improved indefinitely of course depending on what accuracy is required.

In practice we have several problems that limit the accuracy. First, the original data on which all computations are based are experimentally determined numbers and therefore posses a measurement uncertainty. Second, the model which we use is not perfect and so the real physical situation differs from the model in some ways and so represents a further source of uncertainty. Third, the use of uncertain numbers in any computation leads to an uncertainty in the result that in governed by the laws of error propagation.

Generally it may be said that the amount of effort to obtain a further decimal place of accuracy will increase tenfold. It is thus not only scientifically but economically important to decide upon a suitable accuracy as early in an application as possible. Note that not all accuracies are actually possible because they are bounded by the three factors described above.

1.1.5 Certainty

Any statement made about the physical world is only correct with a certain probability. This probability can be very nearly 100% but it will never actually be equal to 100%. A heuristic optimization method will eventually output the best point that it has been able to find. Based on theory, it is often possible to comptue the probability with which this outputted point is the actual global optimum.

Suppose that you are told that this point is the optimal point with 80% probability. Will you be happy? How about 90%? Or 99%? The interpretation of this number is not easy. Having a 90% likelihood that this is the optimal point means that if this algorithm were run 100 times, then statistically speaking you would get this result or worse in 90 cases and a better result in 10 others. Beware that this is a statistical statement, it may well turn out that all 100 trials, if actually performed, would yield worse or equal results.

The question is thus twofold. First, what probability are you happy with? Second, what is there to be done in case the probability is less than that? Clearly other trial optimizations must be performed. But how many and what will happen if the required probability is still not reached? The questions are similar to one we answer with respect to the weather report: How high does the rain probability have to be before one will take an umbrella outside?

In conclusion, we can only say that this is a questions whose answer is emotional and cannot be justified on mathematical grounds. It nevertheless pays to think about it.

1.2 Exact vs. Heuristic Methods

When we ask "what is the optimal point?" we need to already know what kind of answer we are prepared to accept. Most importantly, how accurate does the answer have to be in order to count as a suitable answer? Only the application at hand can provide this information. Is it enough to know that the highest mountain in Europe is in France or do we need its location to within a meter?

In mathematics we distinguish between *exact methods* that deliver a totally definite answer without uncertainty and *heuristic methods* that deliver an approximate answer. In the context of numerical problems, exact methods do not, of course, provide truly exact answers (because real numbers cannot be specified exactly in finite time and space, in general) but rather can provide answers to any degree of specified accuracy desired. The exact method will, however, provide the correct location of the global optimum. Heuristic methods, on the other hand, may get the location of the optimal point totally wrong. Often however, we can know with what likelihood the heuristic method got it right.

Based on this distinction, we would always want to use exact methods. However, the certainty of the best answer has its price: In practical problems, exact methods usually take far too long to terminate to be used. Real life problems usually require heuristic methods to be used.

1.2.1 Exact Methods

The most basic exact method that works always is called *complete enumeration*: List all possibilities and choose the best one. Clearly, this is simple and it will work. However this method is generally impractical as one might see at first sight because the list would take too long to compute in most cases.

The challenge is thus to exclude many possible configurations from an enumeration *a priori*. We might, for instance, break up our solution space into sections and these sections into further sections. We therefore have a hierarchy of sections to search. This is similar to dividing the Earth into continents, countries, regions and so on. We can then ask if we can exclude an entire section from the search *without* searching *in* that section. This would be equivalent to asking if we could say that the highest mountain in Europe is *not* in Germany without looking.

It may seem paradoxical to exclude certain locations from a search without looking there but in mathematical problems, we usually have some knowledge about the problem at hand that allows us to make such inferences *a priori*.

Two particular strategies for using such information are *tabu search* and *branch-and-bound* methods.

When considering solutions to the problem, we always have one current solution and then generate the next one. The difference between them, is a *move*. We thus start with some solution and then move our way through the space of all possible solutions. In tabu search, we archive certain moves on a tabu-list and forbid these moves from being made in the near future. This way, we can avoid the algorithm undoing a previously made move a short time afterwards and thus effectively going in circles. This technique can be made much more complex but it allows effective searching without accidentally moving backwards.

The branch-and-bound technique expects there to be a hierarchy of regions like mentioned above. The branch part of the method creates the hierarchical structure and thus the branches of the search tree. The bound part of the method uses a problem-specific method to compute an upper and lower bound upon the goal function value. Using this bound, we can then decide whether that branch is worth investigating in more detail or not. In this way, we may identify large parts of the solution space as unsuitable for the optimal point without actually looking at the points within that part.

If an exact solution is needed, branch-and-bound is the way to go. Of course, this depends upon a suitable problem-specific branching and bounding algorithm that must be designed for each problem by hand and may require significant research.

1.2.2 Heuristic Methods

Heuristic methods have a chance of delivering a bad answer – finding a point that is good but not the optimum. They have the advantage that they are usually much easier to implement and use and also run much faster than exact methods.

There are two sophisticated and totally general search strategies that are capable of solving *any* optimization problem. These are referred to as simulated annealing (SA) and genetic algorithms (GA). Both have many variants and branch methods. Essentially the idea behind each are the following. In SA we start with a random point and start transiting to other points accepting such a transit always if it improves the objective and with a probability that gradually decreases over time if it does not improve the objective. In GA we start with a family of points that interact somehow to form successive generations of points. Both methods eventually reach a stable state at which time they may be stopped and the best point found is reported.

As a guiding principle, from which we may learn something for our lives, is that both methods first get a general overview of the landscape of the problem. Then the major features of the landscape are explored and determined. Finally, the small details are fixed near the end. Generally, the top-down approach is the correct approach to optimization. The bottom-up approach is not successful. Thus, we recommend to always first get the bird's eye view of a situation.

Both SA and GA are methods that may be applied to any problem. For SA we must define how to transit from one point to another. For GA we must define how two points beget another point for the next generation. The rest of the method for both SA and GA are completely general and may discussed in absence of a particular application. This is the strength of both of these methods.

In this book, we shall prefer SA as the general optimization method. This fact is owed to several practical observations over many projects: (1) SA is easier to implement for a given problem, (2) it makes sense to cut SA off after an *a priori* set amount of time, (3) SA can be shown to converge to the global optimum under very general conditions, (4) SA usually achieves both better results in general and also better results for a certain available time budget, (5) SA only needs to maintain two points in memory at any time whereas the size of the generation in GA must grow with growing problem size. In brief, we prefer SA to GA because of practical performance issues as well as theoretical advantages.

We note in passing that this judgment is not appropriate for *every* problem but merely for most. In mathematics, the advantages of SA vs. GA are complex and not completely understood at present. Practitioners of either method almost always hold to it very strongly and statements of comparison often have an emotional character. When we restrict ourselves to SA instead of presenting both carefully, we do it for the above well meant and well documented reasons as well as the practical reasons that in a book focusing on practical applications a fundamental equanimity has no place as there is simply no room for it in the book and no time for it in the day of industry problem solvers.

In conclusion, SA is good enough for industry work and we recommend it most heartily to all.

1.2.3 Multi-Objective Optimization

In some cases, we may have more than one objective function when seeking an optimum. For instance, if we want to maximize profit and minimize cost or maximize yield while maximizing equipment lifetime, then we have more than one objective and we will have conflicts between them. It is to be expected that there is a point at which any further improvement along one objective creates a detriment to another (such a point is called a *Pareto optimum*, see figure 1.2). This is the challenge of multi-objective optimization. This is, by now, a large research field with many methods. We will discuss the basic ideas here but refer to the literature for more details [116].

The obvious solution is to create a single objective function from the various objectives and so create a new problem that has only one objective function. We may, for instance, translate all objectives into monetary terms and then maximize the financial yield. This is a step that requires human ingenuity to set up a suitable function that really does resolve the conflicts in an acceptable manner.

Such a formulation is not always possible as this involves a solution to all possible conflicts *a priori*. We note that such a formulation used to be popular in economical theory as the sum of all utilities (under the assumption that people act rationally so as to maximize the sum of all utilities). This was rejected in economic theory as unrealistic because humans are not sufficiently rational in that we have preferences that we cannot always express on a sliding numerical scale.

Thus, we are stuck with Pareto optima. The tricky point is that there are usually more than one such points and we must choose among them, see figure 1.2. One way out is to setup a preference hierarchy that disambiguates between various possible Pareto optima, e.g. objective x is more important to me than objective y. For the example in figure 1.2 this is sufficient to uniquely determine the optimal point (the rightmost point on the locus).

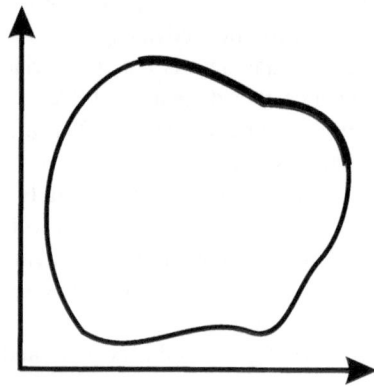

Fig. 1.2 Here both axes describe objectives. The shape describes the locus of possible solutions. Going further along the x axis improves the x objective until we hit the end of the locus. Thus, the rightmost point on the locus is a Pareto optimum as any further improvement of the y objective automatically leads to reduction in the x objective. The highest point in the locus is also a Pareto optimum because any further improvement in the x objective automatically leads to a reduction in the y objective. In this case, all points in the thick line are Pareto optima.

The Pareto approach is particularly useful in social environments where it is important to show that no one's interests will be downgraded. In the event of unclear preference rules, however, there is substantial room for political negotiation. In an industrial setting where we have clearer goals it becomes a little easier but not much. Consider having to rank the following (and more)

1. production volume
2. production losses due to failures
3. equipment lifetime
4. raw material cost and quality
5. final product cost and quality
6. employee safety and satisfaction
7. customer satisfaction
8. greenhous gas emissions
9. environmental pollution
10. corporate image

Due to the fact that many of these preferences cannot be measured accurately and will change unpredictably in the future, setting up a preference hierarchy can be nearly impossible. Should you decide to attempt such a ranking, section 3.3 describes a reasonable method to do so in a group. It is thus likely, in an industrial setting, that one would choose a single financial objective. That is the reason for which we will not go further into this approach.

As a final remark, we must carefully note with Pareto optimization as with any other optimization method in several dimensions: If we are not currently at the optimum, then the path to get there is likely to involve changes in several dimensions simultaneously. In other words, no single remedial measure is likely to achieve the optimum. *Indeed, it is possible that any single remedial measure is going to make the situation worse if it is not accompanied by other measures!*

1.3 Practical Issues

Practically speaking, there are several issues at hand. The most crucial are

1. appreciation that there is potential: This is a management task and is necessary as the trigger of any optimization project. This will be addressed in chapter 3.
2. collecting enough and representative data: Addressed in chapter 4, interacts with the next point and may involve a certain amount of engineering in dealing with instrumentation.
3. understanding the situation: Partially addressed in chapter 5 and presupposes human understanding of the processes involved.
4. making a good model: Addressed in chapter 6.
5. drawing the right conclusions: Addressed in chapter 7.
6. implementing the conclusions sustainably: Addressed in chapter 8 and focuses mainly on the human team.

Many of these issues have to do with the human team and involving them such that the project can be done at all and also that the results of the project actually materialize in real tangible and long-term results. This is an essential point in practical industrial work and that is why we dedicate chapter 8 to this topic.

Supposing that the human aspect is adequately dealt with, we must also address a number of technical issues. The right data must be collected such that it describes the problem as fully as is economical. In environments such as chemical plants or power plants, the instrumentation around the process control system is typically enough for most modeling tasks. In other plants, such as a water treatment plant, the process is significantly less instrumented and we may thus have to decide where to put which sensor.

The data collection process should then occur over a timescale that covers all the interesting phenomena. If, for instance, the seasons play an interesting role (because e.g. the plant is exposed directly to outside temperatures and sunlight), then we would need to collect data over at least one full year. The data must then be cleaned so that only effects that are physical remain in the data, see chapter 4.

Having got the data, we must then make the actual model. This typically requires experts with appropriate tools to create an accurate model in reasonable time.

It should be understood by all involved that most methods that produce an exact solution to any problem are inappropriate here. In an industrial setting we cannot formulate the problem so cleanly or provide data so accurately to even allow the problem to be posed exactly, let alone solved exactly. In addition, the method to produce an exact solution would often require computation times of months or more; to be re-done at the first bug found or the first opinion changed. From the start, we are thus looking for a "good enough" solution. This is the area of heuristic algorithms that produce a good result but not the exact solution. Statistically speaking, the more computing effort put into a heuristic method, the closer the result is to the result we would want. There comes a point where it makes no practical difference anymore and this is the point we need to identify: *How accurate does the answer have to be to satisfactorily solve the problem in the real world?*

This is the principle of diminishing returns in its dual form: Every additional decimal place of accuracy requires more effort than the one before and delivers less actual result than the one before. Thus we must choose well what we need.

Note here that we have four groups of people involved in the process: Management, process related people, project managers and optimization experts. It is the job of management to start the project and provide it with enough importance to get implemented. The process related people must provide the data and live with the conclusions. It is thus essential to excite them about the project. The job of project management is mainly to provide this excitement and also to translate between the process people and the optimization experts as these two groups generally do not speak the same language. The optimization experts are in charge of drawing the right conclusions from the data provided – typically this involves several iterations of asking for more data and more explanations about what the data means.

The process described at the start of this section is not linear from the top down but will most likely circle between the intermediate points several times. It is essential that the project manager keeps the people together and translates between both camps so that it eventually becomes clear "what we have" and "what we want." These two questions will initially be answered incorrectly and will be updated incorrectly several times as well. Only after some time of both parties starting to understand each other will these questions receive a correct answer. This is the natural evolution of such a project and one should not be upset at this. For this reason, it is crucial for the management that started the project to appreciate this point!

1.4 Example Theoretical Problems

For the sake of the discussions in this book, we will usually focus on the traveling salesman problem (TSP) because it is simple and easy to understand, yet provides enough optimization complexity and also room for adding features to it to make it practically interesting.

Classically the TSP consists of N locations and a matrix of distances between them. We then ask for the shortest trip between the locations such that each location is visited exactly once. Thus, we must sort the N locations into an order so that, if we add up the distances between each successive pair on the list, this sum is the smallest possible sum over all such orderings.

A simple computation shows that there are $N!/2$ such orderings. For realistic N, this is a number so large that we could not list all orderings in a reasonable amount of time or space. This makes the TSP into a problem worth thinking about, i.e. we must find the best ordering without looking through all possible orderings.

We can make the TSP into a realistic problem by adding complicating features to it. For instance, some locations may be depots and others drop offs each with a load to be picked up or dropped off. The vehicle that moves between locations may have a finite maximum capacity. The locations may be connected by various roads with differing distances and different toll prices. The driver may be limited due to union contracts in his driving behavior that interacts with speed limits on the roads.

To be clear in our vocabulary, the TSP as formulated above is the problem. If we actually specify the number N and give an actual matrix of distances, then this is an *instance* of the *problem*. This distinction is important as some conclusions apply to the problem as a whole and some only to particular instances.

Another problem would be to find the x such that y is maximal under the condition $y = f(x)$ for some function $f(\cdots)$. Note the principal difference between this and the TSP. For the TSP, it would suffice to list all solutions and check each. This is time consuming but possible. For this second problem, we cannot do this as the number of solutions (all x) is not finite. This is the difference between a discrete problem (such as the TSP) and continuous problem (such as seeking a maximal y).

Chapter 2
Statistical Analysis in Solution Space

" 'You see,' he exclaimed, 'I consider that a man's brain originally is like a little empty attic, and you have to stock it with such furniture as you choose ... It is a mistake to think that that little room has elastic walls and can distend to any extent. Depend upon it, there comes a time when for every addition of knowledge you forget something that you knew before. It is of the highest importance, therefore, not to have useless facts elbowing out the useful ones.' "

Sherlock Holmes [Arthur Conan Doyle, A Study in Scarlet]

In the solution of optimization problems, many factors act in concert to achieve the cumulative effect that we measure using a single cost function. We are dealing with finding a particular microscopic arrangement of many constituent parts – called a *microstate* – in order to attain a desired macroscopic result – called the *macrostate*.

Suppose that you are in a room. This room has many molecules of air that move around in the room. The knowledge of the positions and momenta of all these molecules is the microstate of the room. The macrostate is comprised of a few parameters of interest to you, such as the temperature and pressure of the air. If you were to move a single molecule from one side of the room to the other, would the temperature in the room change perceptibly? No. This observation means that (1) even though a particular microstate leads to a particular macrostate, (2) any one macrostate can potentially be achieved by more than one microstate. The relationship between microstate and macrostate is thus not a one-to-one relationship. By analogy to maps we have one altitude for a specified location but possibly several locations for one specified altitude; as such the location is the microstate and the altitude the macrostate.

The same observation holds true for optimization problems: A particular value for the cost function is usually achieved with many settings of the process parameters. *The optimum state is an exception and is often achieved using only one parameter setting just as the altitude of 8850 meters is achieved only in one location, namely Mount Everest.* In the analysis of the relationship between microstates and macrostates, the analogy to the molecules in the room applies.

P. Bangert (ed.), *Optimization for Industrial Problems*,
DOI 10.1007/978-3-642-24974-7_2, © Springer-Verlag Berlin Heidelberg 2012

As this problem was first investigated by physicists in the context of thermodynamics, the language of the theory uses vocabulary that is reminiscent of thermodynamic processes. This should not be misunderstood as the suggestion that optimization problems are thermodynamic. They are not. The theory that governs thermodynamic processes is, however, so general that it can easily encompass our situation of optimization problems.

The relevant field of physics is called *statistical mechanics*. It derives its name from the fact that the macrostate is essentially a statistical summary of the microstate just as the mean, or average, is a statistical summary of a set of numbers.

In this chapter, we will treat the relationship between microstate and macrostate as developed in statistical mechanics. The vocabulary of thermodynamics will be retained but the ideas will be made sufficiently general that it will become clear how they apply to our situation. For the purposes of this chapter, please suspend any ideas of optimizing. First, we must become clear about how the state of the problem relates to the cost function or, in other words, we must first understand the problem that we are faced with and the answer we desire. Only when this relationship is clear, are we permitted to ask what the state of the problem is that corresponds to a minimum in the cost function.

2.1 Basic Vocabulary of Statistical Mechanics

The energy of a physical system is essentially the same as the cost function in optimization in that nature seeks the configuration of least energy. To understand this from the physical perspective, we quote a description of the concept of *energy* here:

> "Consider a volume of water stationary in a pool at the head of a waterfall. It has what we may call 'privilege of position,' in that once it has dropped over the fall we must do work to return it to its original position. As the water passes over the fall its 'privilege of position' vanishes, but at the same time it acquires *vis viva*, the 'living force' of motion. By passing the water through a turbodynamo, we strip it of its *vis viva* and simultaneously acquire electric power which, vanishing when the dynamo is shorted through a resistance, there gives rise to an evolution of heat. If the water drops directly to the bottom of the fall, without passing through the turbine, *vis viva* disappears without the production of electric power; but at the bottom of the fall the water has a temperature slightly higher than that with which it left the top of the fall – just as though it had received the heat from the above-noted resistor. Now *a priori* there is *no* reason to suppose that 'privilege of motion,' *vis viva*, electric power, and heat – qualitatively apparently utterly different – stand in an relation whatever to each other. Experience, however, teaches us to regard them all as diverse manifestations of a single fundamental potency: energy (Gr. *energos*, active; from *en*, in + *ergon*, work)." [98]

Let us consider a particular instance of an optimization problem. For definiteness, consider a particular instance of the traveling salesman problem. The number of cities and the distances between each city pair is known.

A *microstate* is a complete detailed description of any arrangement of the most basic elements of the problem such that no boundary conditions are violated. Any microstate is thus a *solution* of the problem instance. In the context of the traveling salesman, any ordering of the cities, without repetition, is a microstate and thus a solution in the sense that all such orderings are legal traveling salesman tours. Remember that we are not optimizing yet, we are just describing the problem. If you had an ordering of the cities in which a particular city featured more than once or a city was missing, then this would violate a boundary condition of the problem and thus not be a microstate or solution. In terms of mathematics, a microstate can be expressed as a vector.

A *macrostate* is a global description of a microstate in terms of all the functions that we will later use to optimize the solution. In most optimization contexts the macrostate is the value of the cost function and thus a single number. For the traveling salesman, the macrostate is the total length of the tour.

A *system* is the instance of the problem viewed as an evolutionary entity that changes in time. Mathematically speaking, a system is a series of microstates ordered in time. In the context of thermodynamics, the microstate of the molecules in a room will change from moment to moment in accordance with the laws of physics. In the context of optimization, the microstate of the traveling salesman problem will change from one step of the optimization procedure to the next. In both cases, there is a mechanism of evolution (physical laws or an optimization algorithm) that causes a time-ordered sequence of different microstates. Accumulated from some start time to some end time, this is referred to as a system.

When we have a system, we can take an average of the macrostates over time. That is from the start time to the end time of the system, we select a certain number of macrostates evenly spaced in time and perform an average. The result is called the *time-average* of the system.

Consider again a particular instance of a problem. Imagine now having many copies of this instance. Each copy is put into a random microstate; many will be different from each other but some may be the same. We shall have something to say about the meaning of the word 'random' but will delay it a little. To get a mental picture of this, imagine that the problem consists of a room full of molecules. Now imagine that you have a great many rooms. All the rooms are identical to each other in every aspect except that their microstates – the positions and momenta of the molecules – may be different; as a logical consequence their macrostates may also be different. Each of these copies now evolves over time and thus we have a set of systems. This set of systems is called an *ensemble*. The concept of an ensemble is very important in the treatment of statistical mechanics and thus in our views on the relationship between microstates and macrostates. Please note that we are never going to actually construct an ensemble as this would require too many resources and thus be a practical impossibility. We are just going to consider the existence of an ensemble as a thought-experiment.

At any instant in time, we may record the macrostate of each copy in an ensemble and perform an average over these values. This is called the *ensemble-average*. We can take an ensemble-average at any moment in time including the start time and the end time of the systems in the ensemble. If the value of the ensemble-average does not change with the time at which the average is taken (with the possible exception of some initial time period), the ensemble is called *stationary*. Physically, this is usually called *equilibrium*. Note that if an ensemble is stationary, the many possible ensemble-averages differing due to their start and end times all take the same value and thus there is in fact only one ensemble-average value. Stationarity is thus a crucial concept for us to speak of *the* ensemble-average as opposed *an* ensemble-average.

Having discussed two averaging procedures, the time and ensemble averages, it is interesting to look at how they differ. In both averages, we list the macrostates of a large number of microstates and perform an average. If the number of microstates is sufficiently large, then the averaging process itself should be stable and the results represent truly underlying differences. In the time case, the microstates are connected by the evolutionary laws of the process (physics or an optimization algorithm). In the ensemble case, the microstates are connected by their initial selection and then their evolution according to the same laws. If an ensemble is stationary and the ensemble-average is equal to the time-average, then the ensemble is called *ergodic*.

To be clear, ergodicity is a good thing. We like ergodic ensembles. Situations where ergodicity is not valid are generally very hairy indeed. The reason for ergodicity being desirable is that if an ensemble is ergodic, we can replace the time-average by the ensemble-average in any mathematics that we will want to do. This is an elemental difference due to the fact that computing a time-average would require the solution of the time-dependent partial differential equations that govern the evolutionary laws of the process. We do not like doing this. Respectively, in many situation we cannot do this. Computing the ensemble-average is relatively easy due to the fact that the individual copies are randomly assigned a microstate and the evolution in time does not play a role (the ensemble is stationary). To perform such an average, we merely need to generate a lot of random microstates, take our average and the deed is done. Computationally speaking, we actually create these many microstates in the computer. Doing this, including the ensuing taking of the average, is a simple presentation of a collection of techniques commonly called *Monte-Carlo computation*. As before, we delay the definition of the word 'random.'

In keeping with the language of statistical mechanics, we are going to use the word *energy* as a synonym for the objective or cost function of the optimization. Physics is effectively one big optimization problem as physics postulates that nature always evolves in order to minimize its energy. Recalling our above definitions, energy is effectively the number representing the macrostate. As every microstate has one corresponding macrostate, we can associate an energy with each microstate.

At this point in the discussion, we are going to create our first basic assumption, namely: The number of possible microstates is finite. Please note that in general the number of microstates is very very large but we demand that it not be infinite. This is important because we want to start counting how many microstates belong to any given macrostate and we want these numbers to be finite so that we can do arithmetic with them. As the number of microstates is finite, we can label them with an integer. The order does not matter for this purpose. We will denote the energy of microstate i by E_i.

The only thing left in the presentation is to be clear about the term 'random.' We will make our second basic assumption: The probability of the system being in any one microstate is equal to that of any other microstate. If there are N microstates in total, then the probability of the system being in microstate i is $P_i' = 1/N$. It is now easy to create an ensemble. We simply select microstates from the set of all microstates each with probability $1/N$. Due to this procedure, we will get an ensemble, we will be able to compute an ensemble average and, if the ensemble is ergodic, this will be equal to the time average and thus give us something interesting.

The probability of the microstate was thus settled by assumption. But what is the probability of the associated macrostate? Well it is simply the number of microstates associated with this macrostate divided by the total number of microstates, $P_i = N_i/N$. While this is an easy formula, it is far from easy to work it out as we, in general, will be hard pressed to compute N_i. Thus, we must find a formulation that is easier to compute.

To discover this, we first talk about temperature. Going back to the physical case of the room full of molecules, we note that this room does not actually exist in isolation but rather it is part of the world and exchanges energy with the world. After some time, so our experience tells us, the temperature in the room will equal the temperature of the world. In statistical mechanics, the world is therefore referred to as a *heat bath*. The concept of *temperature* enters our discussion here as a crucial parameter that is supplied by the external forces that act upon our system; see section 2.4 for this concept. Also, we will assume that we know what the temperature is because we can measure it in the heat bath. We will find that the concept of temperature will play a major role in our later optimization efforts. It should be understood however again that while we are using vocabulary from statistical mechanics, the concepts are much more general and can be applied to non-physical systems. Temperature, for example, is just a macroscopic parameter of the system supplied by the external heat bath forces that govern the system evolution.

Now that we know what temperature is, in statistical mechanics, it is possible to derive what P_i is actually equal to. We will not follow the derivation here as we are concerned only with the interpretation of these results. We have what is called the *Maxwell-Boltzmann distribution*,

$$P_i = \frac{g_i e^{-E_i/kT}}{\sum_j g_j e^{-E_j/kT}} \tag{2.1}$$

where T is the temperature, k a constant known as the *Boltzmann constant* and g_i is the *occupation number* of the energy E_i, i.e. the number of microstates having energy E_i. The denominator of the distribution is referred to as the *partition function* and serves several important uses in statistical mechanics to the extent that complete knowledge of the partition function essentially means complete knowledge about the system – at least with regard to all the things that physics is usually interested in, i.e. the macroscopic description of the system. The partition function cannot practically be evaluated as defined because it is a sum over all microstates and the number of microstates is very large indeed. Supposing that we could write the partition function in a way to be able to directly evaluate it, we could perform ensemble-averages and thus time-averages.

With the partition function

$$Z = \sum_j g_j e^{-E_j/kT}$$

we may find a number of other crucial thermodynamic concepts such as the energy E, the entropy S or the Helmholtz free energy A,

$$E = -\left[\frac{d\ln Z}{d\beta}\right]_v = kT^2 \left[\frac{d\ln Z}{dT}\right]_v,$$

$$S = \frac{d}{dT}\left[kT\ln Z\right]_v,$$

$$A = E - TS = -kT\ln Z$$

where $\beta = 1/(kT)$ and the subscript v indicates that the derivative is to be taken at constant volume. Thus, our knowledge of thermodynamic properties of a particular system is limited by our ability to compute its partition function!

Note that if two microstates have the same energy, their contributions to the sum in the partition function are the same. This is a desirable property as two microstates of the same energy would belong to the same macrostate and should therefore be, macroscopically speaking, indistinguishable. Therefore, it is good that their microscopic contributions are the same.

While the Maxwell-Boltzmann distribution works very well for certain systems in the physical world, it is not necessarily true for all physical systems or indeed for non-physical systems. An abstract optimization problem can be profitably analyzed using the language of statistical mechanics but we must remember which conclusions of statistical mechanics are of a generic nature and which apply particularly to specific elements of physical nature.

2.2 Postulates of the Theory

It is possible to build up statistical mechanics as a formal theory based on axioms. This fact is important beyond making the theory formally clean because it shows the fact that the theory is very generic and applies to many situations that have nothing whatsoever to do with thermodynamics. The postulates are these six [78]:

1. The constituents of the system obey certain laws of motion that themselves do not change. In the thermodynamic context, these are classical or quantum laws of physics. In the industrial context it will usually be classical physics only.
2. An observation is a simultaneous and instantaneous measurement of a set of indicators, which each take the value zero or one only. The instants at which these observations may be made are discrete and equally spaced.

3. Observations cause no visible disturbances in the macrostate of the system under observation[1].
4. The successive observational states as given by the indicator values form a Markov chain.
5. Any microstate may be the initial state.
6. A system with finite energy has finitely many microstates available to it.

All of these postulates are quite clear and simple for a problem that is well defined. The interesting postulate is the fourth concerning the Markov chain. Effectively this means that the system has no memory of its previous states as a Markov chain is defined by a transition probability matrix in which the probability of the future state depends only upon the identity of the current state and not the past states. A Markov process is thus a probabilitistic (stochastic) process without memory.

It is clear that this assumption is not strictly true about every system in nature but it is close enough to being true that the theory leads to interesting results about nature. As we are concerned with optimization in this book, however, we need to ask ourselves whether the optimization method will respect this and the other postulates. If it does, then the following theory will apply to the analysis of its results. In general the methods that are used for optimization do respect the Markov postulate and we may thus proceed.

Through this postulate, we effectively take the probability of a certain observational state to be an intrinsic characteristic and we implicitly assume that this is measurable for example by repeating an experiment several times[2]. The probability of an observational state thus takes on an ontological value similar to that of an object's mass in classical physics.

While the microstates obey physical laws and are thus deterministic[3], the observational state, i.e. the macrostate given by the indicators, does not change deterministically. We want a reliable statistical description of its evolution – that is the purpose of the theory.

Why is that? Note that it is very complex to observe the microstate at all times and to model it deterministically. It is far more expedient to model the macrostate because it has few parameters and we are interested in it. Due to the fact that it does not change deterministically, all we can expect to receive therefore is a statistical description of the macrostate. In order words, while we can say that a certain microstate will *definitely* transit into another specific microstate, we can only say this *with a certain probability* in the context of a macrostate.

From these postulates, it is possible to derive all other statements in statistical mechanics including the famous four laws of thermodynamics, which are the following.

---

[1] From quantum theory we know that this fundamentally wrong but we are dealing with systems far larger than quantum systems and we may thus reasonably assume this. Please do note that there is a large debate about the role of the observer on a physical system and that this point is an assumption and not a statement of fact.

[2] The concept of probability will be further discussed in chapter 5.

[3] We will not concern ourselves with the nature of determinism in quantum mechanics as we are not dealing with the application of this theory to physical situations.

0 If systems A and B are each in equilibrium with system C, then A and B are also in equilibrium with each other.
1 The energy of an isolated system is conserved.
2 The entropy of an isolated system may not decrease.
3 A system cannot be brought to zero absolute temperature.

The concept of energy was already explained. In the following, we will explain entropy and temperature in more detail.

2.3 Entropy

Having covered some basic concepts, we will turn to one of the most central concepts of statistical mechanics, the *entropy* of a system.

"The first principle of thermodynamics poses the concept of 'energy'; the second principle, the concept 'entropy.' Feeling that we know what energy *is*, we demand to know what entropy *is*. But now, in point of fact, *do* we really know what energy is? The classical dichotomy is matter vs. energy, and energy may then be defined as whatever produces heat. But in the early 20th century this dichotomy was undermined by recognition of the interconvertibility of mass and energy, and to the question 'What is energy?' we can now give only the unsatisfactory reply 'It is *everything*.' Yet however great may be our uncertainty about the intrinsic nature of energy, the thermodynamic significance of that concept remains wholly unimpaired. ... Indeed we don't need to know what energy is, but we do find it satisfying and instructive to use the kinetic-molecular theory to *interpret* internal energy in terms of the kinetic and potential energies of atoms and molecules. Neither need we know what entropy is, but we find it satisfying and instructive to use the kinetic-molecular hypothesis to *interpret* entropy in terms of the 'randomness' with which atoms and molecules are distributed in space and in energy states. A simple illustration of the subtle concept of randomness is found in the ... example of a bullet abruptly stopped by a sheet of armor plate. The bullet's gross kinetic energy disappears, and in its place appears thermal energy that manifests itself in a rise of temperature. *Before* the impact, all the lead atoms comprising the bullet traveled together, as a unit, because all had a single directed component of motion superposed on their uncoordinated thermal motions. *After* the impact, this directed component is randomized: when the bullet's gross motion vanishes, the constituent atoms acquire an increased energy of random thermal motion, which is reflected in the temperature rise. Observe that this molecular picture renders easily intelligible the striking disparity of the following two cases: (i) a moving bullet, when stopped, becomes hotter; and (ii) a stopped bullet, when heated, is *not* thereby set in motion. This otherwise puzzling asymmetry or unidirectionality grows out of a statistical situation amply familiar in everyday experience. Consider for example that a new deck of cards, factory-packed in a regular arrangement of suits and denominations, is soon randomized by shuffling; but we think it highly improbable that, by further shuffling, we will soon return the pack to its original highly-ordered arrangement." [98]

Consider a closed plastic bottle of water. If you squeeze it, the level of the water will rise in the bottle. If you let go, the level will sink back down to its former position. The squeezing is therefore what is called a *reversible process*. Smashing a glass onto the floor and seeing it break into pieces is called a *non-reversible process*. The difference between them lies in the energy that you have to put into reversing the process. Clearly you can mend a broken glass – but only with effort. Restoring the squeezed bottle to its former state does not require exchange of energy between the system (bottle) and the external world (your hand).

In the real physical world, no action is truly reversible as there is always friction. For instance, when squeezing the bottle, you are actually transferring a small amount of energy to the water which manifests in a rise in temperature. In practical terms, this does not matter because the temperature increase is small but it is there nonetheless and so the process is not quite reversible. However, it is clear that the breaking of the glass is a lot less reversible than the squeezing of the bottle.

Thus, we ask ourselves for a measure of reversibility. This measure will be called entropy.

Suppose that the letter A labels a particular macrostate and $\Omega(A)$ denotes the number of microstates giving rise to that macrostate. Then we define the *Boltzmann entropy* to be $S(A) = k \ln \Omega(A)$ where k is a universal constant known as the *Boltzmann constant* the numerical value of which must be determined by experiments. This definition gives rise to the desirable fact that the entropy is additive. This means that if we make a larger system C by combining two systems A and B, then we have $S(C) = S(A) + S(B)$. While this is not a fundamental requirement of the universe, it is desirable because it makes life easier when computing and also simply makes sense that a system property would add when systems are added. Please note however that this is a definition and not the result of any argument!

The same definition leads to the fact that if macrostate A transits to a different macrostate A', then there could be a change in the entropy of $\Delta S = S(A') - S(A) = k \ln(\Omega(A')/\Omega(A))$. Note that, by the laws of logarithms, this change in the entropy can be both negative, zero or positive.

The second law of thermodynamics states that entropy must not decrease. However, this statement applies to an isolated system and not to a system that has energetic contact with other systems. In the physical world, true isolation is not possible but we can get very close in carefully constructed circumstances.

Let us consider the meaning of changes in entropy for a moment. If the entropy increases, this means that the number of microstates for the observed macrostate has also increased. Macrostates that have a larger number of microstates are more likely to be observed by the fundamental postulate (and please note that this is a postulate) that all microstates are equally likely to occur. Thus entropy increasing means that the system moves to a more likely state. We have a special term for the macrostate that has the highest probability of being observed, we call it the *equilibrium* of the system.

By contrast, if the entropy decreases, the system moves to less likely states. While this is of course allowed, it is by definition unlikely to happen. In colloquial terms, the effort that must be expended to get a system into a less likely state must come from outside the system and will incur an energetic cost in the outside that will lead to an increase in the entropy there. For example, the glass that broke when we dropped it can be fixed by effort expended by us on the system thereby increasing our body's entropy.

Practically, if we could plot the entropy over time of a real system, we would thus see small ups and downs everywhere from natural fluctuations but we should see a global trend upwards towards equilibrium. If this cannot be seen, then something happened: An external action was made that influenced the system towards a less likely state. A state of particularly low likelihood is the *ground state* of the system. This is the state of least energy in the physical world or the least cost function value in the computational world. Generally speaking, the ground state has very few microstates associated with it and often only a single one. Thus this state is very unlikely to be observed naturally. Yet, this is the state we wish to find when doing optimization calculations. In the context of optimization, the algorithm is the external world influencing the system and thus it is the algorithm that constitutes the heat bath.

Therefore, when plotting the entropy over time of a search for the ground state, we should see it decrease. It will generally not do so uniformly but rather have distinct local peaks and even discontinuities associated with it. These discontinuities are very distinctive and interesting features of the evolution because they are the telltale signs of so called *phase transitions*. In the physical world, a phase transition is the change from water to vapor or water to ice and vice versa where the phases are generally solid, liquid, gas and plasma. The ground state will generally be a solid state but we may have to begin searching for it with a system in the gaseous state and thus undergo two phase transitions (to liquid and then to solid). To make things more interesting, there are phase transitions of a more subtle nature, called *second-order phase transitions*, which are not visible in the entropy itself but rather in its first derivative. Formally speaking, a phase transition is a discontinuity in the entropy over time and a second order phase transition is a discontinuity in the derivative of entropy with respect to time.

To be clear, if we want to bring a physical system from a gaseous state to a solid state, we must cross two phase transitions. In cooling the system, we must be very careful at these points because local freezing may set in that will later make it impossible to reach the ground state without re-heating the system. Thus, we must cool very slowly at these points and hence there is a necessity to know where the phase transitions occur.

It is widely documented that phase transitions are observed in computational systems as well as natural ones. These are important events that must be negotiated carefully by the optimization algorithm "cooling" the system because systemic damages can occur during the change of phase. To illustrate this point, consider the freezing of a glass of water. Typically, the boundary will freeze first and slowly the freezing process will permeate to the center of the bit of water. If this occurs in the context of a fragile substance, such as a crystal, the parts that are on the boundary between the frozen and liquid portions experience a stress that may cause local damages in the crystal structure. These damages then freeze and so there is not enough energy anymore to repair the damages by natural fluctuations. This is the so called *freezing in* of local structures. Once this has been done, finding the true ground state is impossible without first heating the system up again and thus allowing the defect to be repaired. The same effect has been observed many times in combinatorial tasks and so we must beware of phase transitions that will block the path to optimality if they are not negotiated carefully.

So what does negotiating carefully mean? It means two things. First, we must allow the system to cool (i.e. loose energy to the outside world) as slowly as possible so that local stresses are small and damages unlikely. Second, we must apply the cooling as uniformly across the spatial extent of the system as possible so that the the glass of water might freeze as a unit and not from the outside inwards. While being very difficult in physical terms, the second can be achieved more easily in a computational environment.

2.4 Temperature

The concept of temperature is fundamental to statistical mechanics as it is a major macroscopic state variable and closely related to the concept of equilibrium between two systems. The definition of *equilibrium* is[4]: Two systems in the thermodynamic context are in equilibrium with each other if and only if they share the same temperature. It makes little sense to speak of the temperature of a system that is far away from equilibrium as it would then be impractical to determine its temperature. Note that a physical temperature is measured by a thermometer (one system) being put into a substance (another system) and these two systems must be allowed to come to an equilibrium before it is sensible to take a reading of the temperature.

When we speak of something being hot, we mean the subjective impression that the object is transferring a lot of thermal energy to us and that we are experiencing a change in our state of being as a result. When you touch a hot stove plate, you burn your finger – a lot of energy has been transferred causing an increase in your body's entropy. This indicates a high temperature. When you touch an ice cube, some energy is removed from you and lowers your body's entropy as you move to a less probably state. This indicates a low temperature.

In terms of nature, temperature is thus a measure of the molecular agitation of some substance. If the molecules are more agitated, then the temperature is higher. Generally this also means that the pressure increases and thus the body will expand if it is allowed to do so leading to the well observed fact that objects get larger as they get hotter.

The *Temperature*, denoted by T, is defined by the derivative of the energy with respect to the entropy, $T = dE/dS$. Note that temperature is thus a defined concept in terms of the two basic concepts of statistical mechanics: energy and entropy.

We may ask how much heat (or energy) we must supply to a substance in order to increase its temperature by one unit. This is called the heat capacity of the substance. The heat capacity per unit of mass is the *specific heat capacity* or simply specific

---

[4] This is the definition of equilibrium *between* two systems. What concerns the equilibrium *of* a system, we have already encountered it: Equilibrium is the macrostate corresponding to the most microstates, i.e. the most likely macrostate.

heat of that substance. It is an important concept because it effectively translates between the concepts of energy and temperature.

The specific heat is a characteristic of a particular substance but it is not a constant. In fact, it depends upon temperature, volume and pressure. As the temperature gets large, the specific heat is approximately constant. The specific heat is particularly interesting as we approach absolute zero temperature or the ground state of the system. In nature, it follows *Debye's law* from quantum theory. Here, we have

$$c_v = 9Nk \left(\frac{T}{T_D} \right)^3 \int_0^{T_D/T} \frac{x^4 e^x}{(e^x - 1)^2} dx$$

where N is the volume (effectively a value needed for the physical value of c_v which we may ignore for the purposes of optimization), k is Boltzmann's constant (also a constant that may be ignored for optimization purposes) and T_D is the Debye temperature. The Debye temperature is a material property. For the purposes of optimization theory it may be estimated as that temperature where the specific heat (in the direction of lowering the temperature starting from a high temperature) is first observed to decrease significantly below an initially approximately constant value. In figure 2.1, we display the typical evolution of the specific heat according to the Debye model as compared to the related Einstein model. This is what we would expect to see in the evolution of an optimization problem and we may use this to interpret our progress.

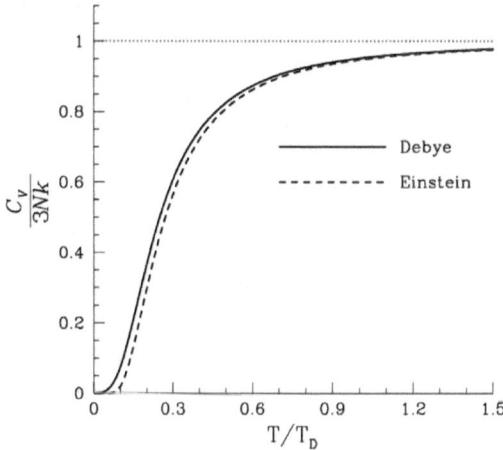

Fig. 2.1 The evolution of specific heat according to the Debye model as compared to the related Einstein model. We use this model to gauge our optimization progress while measuring the specific heat of our currently proposed problem solution.

The reason for studying specific heat in an optimization context is that it allows us to track our progress towards the ground state of the system to be optimized.

We may define two kinds of *specific heat*. The *specific heat at constant volume* c_v and the *specific heat at constant pressure* c_p are defined to be

$$c_v = T \left(\frac{\partial S}{\partial T} \right)_v ,$$

$$c_p = T \left(\frac{\partial S}{\partial T} \right)_p$$

where the differential in both cases is made subject to the requirement that either the volume or the pressure respectively must remain constant.

2.5 Ergodicity

One of the most intriguing concepts in statistical mechanics is that of ergodicity. It is related, as discussed earlier, to the relationship between time-averages and ensemble-averages.

In statistical mechanics, we are primarily interested in the equilibrium state for a particular macrostate. We will thus want to know the value of some interesting quantity $G(\alpha)$, a function of the microstate α, in the equilibrium state and denote its value here by G_{eq}. As the system always tends to the equilibrium state, this value is equal to the time average of G for a long time, i.e.

$$G_{eq} = \lim_{t \to \infty} \frac{1}{t} \int_0^t G(\alpha_{t'}) dt'.$$

The limit in time poses a practical problem. We cannot measure or compute such a limit in practice and so must look for an alternative means by which to obtain this result.

There is a related concept, which takes an average over local phase space (phase space is the set of all microstates), i.e. over microstates with a similar energy. We thus get

$$G_{ps} = \lim_{\Delta E \to 0} \frac{\int_{\omega(E)} G(\alpha) d\alpha}{\int_{\omega(E)} d\alpha}$$

where $\omega(E)$ is that region of phase space, i.e. the set of microstates with energies in $[E - \Delta E, E]$. This phase space average is thus an ensemble average. This is something that we can compute and measure.

We have restricted ourselves to a neighborhood in energy and energy is always an invariant of the motion[5]. If the energy, and functions of the energy, is the only invariant of the motion, the phase space is called *ergodic*. If there are other invariants, then the above phase space integral must be restricted to neighborhoods of these other invariants around the value of the current macrostate also.

The *ergodicity theorem* now states that if phase space is ergodic, then $G_{eq} = G_{ps}$.

This theorem allows us to replace something that we want to know but cannot measure or compute with something else that we can measure and compute. Thus it is very important to know whether phase space is ergodic or not in any particular case.

Let us analyze the situation by focusing on the concept of a Markov chain, which governs the evolution of the system according to our set of postulates. Recall that a Markov chain is a series of microstates where each microstate is arrived at from its predecessor in such a way that the probability to obtain any given microstate depends only upon the present microstate and not upon the history of previous microstates. Such a chain therefore has no memory and is thus particularly easy to model.

An observational state in our Markov chain is called *transient* if there is another state that the system can reach from this one but the system cannot return[6]. A state that is not transient is called *persistent*[7]. The persistent states can now be grouped such that two persistent states will belong to the same group if and only if they can be reached from each other. These groups are called *ergodic sets*. Essentially ergodic sets imply that we may more or less freely move between states within the same ergodic set but are effectively forbidden from going to another ergodic set.

Above, we spoke about the energy being an invariant of the motion and we thus having to restrict attention on states within an energy neighborhood. This is another way of saying that we must focus on one ergodic set of states. If we have more than one ergodic set, we must focus on one of them in order to perform our phase space integral and have it equal the time average of whatever value we are interested in.

So far, things are clean. However there can be a problem. We have differentiated transient and persistent states by the property of being able to return to them. As such the definition is precise. In practice, what also matters is how many transitions (i.e. how much time) are necessary for an eventual return. Sometimes it is quick to return and sometimes a return is possible only after a great many transitions. States

---

[5] The phrase "invariant of the motion" means that the value does not change over time. In the case of energy, this will not change as we have a law stating that energy is conserved.

[6] An example of a transient state is a hot cup of coffee at room temperature. This will naturally tend to cool down and so it can reach this cooler state. However, it will not be able to return to its hot state – unless it is acted upon by an external system such as a microwave oven.

[7] An example of a persistent state is that of a cup of coffee at the same temperature as the room in which it is located. This state may occasionally transit to other states but can and will return to this state of equilibrium.

that are persistent in principle but only after a time that is longer than our typical observational time period are called *pseudo-persistent*[8].

These will cause the splitting of an ergodic set into several subsets that are each an ergodic set for the realistic time-scale defined by our observational period. The existence of this effect is known as *ergodicity breaking* and represents a major computational problem. The problem has several features: (1) It is hard to know what states are pseudo-persistent in advance, (2) it is hard to diagnose ergodicity breaking when it happens and (3) the inherently long times necessary for a tour around the ergodic set increases the time needed for a reliable computation. During the use of an optimization algorithm, we start somewhere and then move from microstate to microstate until we believe to have found the optimum. This moving process could get stuck in one these pseudo-persistent areas and thus practically prevent our algorithm from exploring other areas. If the true optimum is in that other area, we are unlikely to find it. In the language of optimization, these points are called local minima (of sufficient depth and width to limit our evolution from going away for the observational duration).

Particularly for optimization purposes it is troublesome if we spend a very long time in a restricted section of the ergodic set without exploring the rest of it as it could be that the optimum we are looking for is in that rest. To have reasonable confidence that we will find the optimum, we must therefore increase the observational period. However, the effect of ergodicity breaking can occur on several time-scales and so the period may have to be increased by an impractical amount. Moreover, we cannot know how much we need to increase it by unless we can detect what is going on. In short, ergodicity breaking is a major stumbling block to efficient optimum finding and a response to it needs to be found.

What are appropriate responses? A very general strategy is called *restarting* in which we execute the optimization algorithm several times from different randomly selected starting points in the hope to get into all the pseudo-ergodic sets at least once. In fact, this is the response of choice in the field for a variety of optimization algorithms. Uniformly over the entire space of possible solutions, we select N of them and start a full optimization from these points. To save time, these N optimizations can be run in parallel as they do not interfere at all. We then take the best answer. It is observed that the answer quality improves approximately logarithmically. There is thus a law of diminishing returns as we crank up the effort put into a problem's solution. It is also observed that, for relatively low N, the gain is of the order of a few percentage points and so, in general, substantial enough to be worth the effort. We highly recommend augmenting any optimization algorithm with this simple method.

---

[8] A pseudo-persistent state is any state that takes so long to change that we are in danger of not seeing the change in our observational period. An example is the heating of water. If we supply heat to a large quantity of water, it takes a long time (respectively a lot of heat) to create even a small rise in temperature. Another example is a lake at high altitude. The water should flow down to reach a lower energy state but it is limited by the mountain. Eventually erosion will bring the mountain and thus the lake down but this takes a very long time.

Chapter 3
Project Management

In practice, we find that industrial optimization projects typically involve several roles of people. These roles may be filled with more than one person each and these people may belong to different legal organizations. The roles generally are the following

1. User Manager: This role actually orders the project to be started, obtains the budget and finally accepts the deliverables. This role may include persons from technical management, purchasing and organizational management of the organization interested in using the project's results.
2. User Helper: This role helps the project during its execution but will not directly use the project's results when the project is finished. This may include various experts from production and research in the organization wanting to use the project's results.
3. User: This role is going to actually use the project's results and may or may not be involved in the project itself.
4. Implementation Manager: This role is the management role of the organization that will perform the project. This involves people from economic and legal departments making sure that the project can actually be completed successfully.
5. Implementor: This role comprises all people that actually work on the project itself. This may be populated by persons from both organizations and so the dividing line between this role and the User Helper seems fuzzy. The division occurs in that User Helpers participate occasionally when their expertise is needed and the Implementor is essentially a full-time participant in the project.
6. Project Leader: This role directs the evolution of the project. This role should be populated by at least two people, one representing the three user roles and one representing the two implementation roles.

As we are talking about *industrial* projects, these roles generally involve a number of individuals. These have needs and wishes that are difficult to fulfill and even express. To this end, there are several project management suggestions that we will discuss here.

P. Bangert (ed.), *Optimization for Industrial Problems*,
DOI 10.1007/978-3-642-24974-7_3, © Springer-Verlag Berlin Heidelberg 2012

First, we will discuss the actual management philosophy between all the people in an environment of uncertain needs. Second, we will discuss how the process of obtaining the required data can be made efficient. Third, we will discuss a method by which goals may be prioritized and conflicts resolved.

3.1 Waterfall Model vs. Agile Model

For many years, project management was dominated by the so called *waterfall model*. This model compared a project to the production of an automobile in which each action could be done separately from the others in a sequence in which the second action could only start once the first had been completed. Basically, we count a few major steps along the waterfall road:

1. Requirements definition
2. Analysis of requirements to see which pieces were not yet available
3. Design of missing pieces
4. Implementation of the designed pieces
5. Assembly of all pieces
6. Test
7. Production

In many industrial optimization projects, we have observed what many people have observed in other contexts, particularly in the IT industry. The shortfalls of the waterfall model occur mainly in its linear orientation.

In the stage of defining the requirements, the user organization is generally not able to formulate the requirements in full and in sufficient clarity. A vision and a wish is present but it is amorphous at this time. If these requirements are truly established as the basis of a project, then this project will most likely deliver a result that no one wanted and no one will use. In the worst case, the project will do harm. At the very least, much money and effort must be spent in repairing the differences between what has been done and what is needed.

When it is discovered that the project has failed, the cycle is repeated. This will often yield similar results because again requirements were not precisely and fully defined. We learn in practice that users' requirements change (1) in time, (2) in discussions with others, (3) when seeing some results, (4) based on their mood of the day and so on. Then again, the user organization includes diverse and many people often with conflicting wishes even when they can formulate them. Nailing an organization down on a particular definitive wish list is a practical impossibility at least before the project has started.

We now derive a project management suggestion of our own. This is an evolution and combination of three methods that are called *agile* [118], *scrum* [114] and *kanban* [99]. In brief, the features are of iterative project evolution, much communication and the acceptance of imperfect and impermanent project goals by all concerned.

We begin with a definition of the problem that the project is to solve and a vision of the project's solution. Note that we are not defining requirements. We are defining what is wrong and how the world would look like if it were no longer wrong.

Based on this vision, we design an intentionally simple mini-solution that captures the vision but includes no features of any comfort whatsoever. Discussions of ergonomics, look-and-feel and so on are totally ignored at this stage. The focus here is to capture the problem itself and to make the vision tangible.

This is delivered to the user organization and is modified together with the user organization until this mini-solution actually captures the problem. During this discussion, we will learn that even the initial definition of the problem was incomplete. At the end of this, we should have something that truly captures the problem.

Now we can work on capturing the solution. We do this in the same manner as for the problem. Completely ignore the mechanism of going from problem to solution at this point – if you do not know how the solution looks like, it is a waste of time to design the method to obtain it. The user organization will only discover how it wants its solution to look like after an iterative process of checking out several suggested variants.

Now we can work on the actual obtaining of the solution from the problem. We will implement a simple variant of this only at this point. The aim is to arrive at a prototype system as quickly as possible. It is a mistake to get too fancy at this point, it will just be modified later. Just get something working.

At this point, we have a prototype. This should now be installed in the industrial plant and from now on investigated on the live object. We need to demonstrate the successes and problems of the prototype in its real environment even though we know it is only a prototype. As a rule of thumb, this event should not occur more than three months after the start of the entire project. If we take longer than this, the project has fundamental flaws and might have to be reorganized or the prototype is not simple enough and must be stripped down.

We will now find that several requirements will enter our definition that were never before thought about. The interaction with the real plant will yield constraints that are present but forgotten by all concerned. There may be compatibility issues, timing issues and so on. The prototype will be modified now until it actually works stably in the real plant – that is to say that it delivers a solution in such a manner that this solution can be used and is actually useful.

For the first time do we now ask the questions: Does the solution quality need to increase or is it already good enough? Does the solution require additional features to be more "comfortable" or not? Often we find that the prototype is very close to the final solution and all the fancy buttons initially desired are no longer wanted because people have seen that the simple system delivers.

We see the basic ideas are (1) to focus on solving the problem without any frills, (2) to get something working as soon as possible, and (3) to involve the user organization in all steps.

The very first step is, of course, to educate the user organization about this manner of project management and to explain that this saves a great deal of cost and time. Only then is there enough motivation to be constantly involved. Many people in management would like to order a project, wait and then have it delivered. It is important to observe that this is almost impossible (without conflict upon delivery) and will certainly cost more.

So far, we have described the process from the point of view of interaction between the user and implementor organizations. This is called the agile paradigm[1].

We note in passing that these steps and meetings should be carefully documented and the minutes to all meetings should be signed by responsible people from all involved organizations. This makes sure that people can "remember" what they have said in the past and adhere to certain decisions that cannot easily be modified. It also makes sure that the project remains on track economically.

From the point of view of the implementors we need to manage the constant process of change. The implementors are viewed as a scrum, that is to say a rugby team. The terminology is borrowed from the IT industry. It refers to a group of people working on the project. There is a leader of this team but this team also has a so called scrum master different from the project manager. The scrum master's job is to make the scrum's work possible. The scrum master is to eliminate obstacles and as such generally refers to the implementing organization's management to clear up economic, legal and other hindrances to the actual team being able to do what it must.

The scrum meets daily to discuss what has been done the day before, what will be done on the present day and what the problems are that may prevent or delay this. The first two points are resolved by the project manager. The last point is to be resolved by the scrum master. The project manager then assigns tasks to people and makes sure these are done.

It is a key element to the agile/scrum combination that the work be divided into small chunks. If we meet a large chunk, we have to investigate ways of splitting it up into several small chunks. Ideally a chunk has the size of one person's work for a period of 2 – 5 days. If the chunk is smaller, then the management overhead for assigning and checking the steps is too high. If the chunk is larger, then the individual is in danger of implementing into a cul-de-sac and wasting time.

The scrum approach is heavy on testing. Each element should be tested before it is considered done. Ideally, tests are made at first by the implementing individual first and then by someone else again. In a software environment, testing may be automated by writing a testing program. The chunk time of 2 – 5 days includes the time to test the chunk's correct completion.

The chunks currently in progress are displayed publicly in a so called burndown chart. This is most often a whiteboard or a pin board. Here we may see what is, has been and will be done on the project.

This is also the point at which the third element, kanban, will enter our methodology. Each member in the scrum will have particular qualifications and specialties. Any task, a so called ticket, will pass through various people's hands and pass vari-

---

[1] Certain features of the agile paradigm as understood in the IT industry such as evenly spaced meetings with the user organization and so on seem to be too restrictive for industrial projects that interact with the plant. Here we understand agile to refer to many small steps taken together with the user focusing purely on the problem's solution before any luxury features are considered.

ous virtual stages. Such stages may for instance be: specification, research, design, implementation, test, verification, packaging (into the project's whole). These stages may be done by various people or not. The identity of the stages must be altered to match the project's needs.

Every ticket will be tracked across the stages. There will be a limit on the tickets per stage at any time. A ticket is "pulled" by a person who has complete a task and who has capacity to take on another ticket. Thus, we can easily see where tickets accumulate and may react in a timely manner if delays occur. This process is displayed on the above mentioned board and is maintained by the scrum itself. The project manager helps to assign tickets to people and observes problems; the scrum master resolves over-arching difficulties such as the number of people in the scrum, the available resources and so on. In this way, we also incorporate the aim of continuously improving the process (kaizen) via the interaction of scrum and management via the kanban board.

It is advisable that each ticket receive: (1) a priority, (2) an estimation of the effort required for completion, (3) a dependency note on other tickets, if applicable, (4) a deadline, (5) team role who is to fulfill the ticket. During execution, the ticket should receive information as to what has been done to it.

Many presentations of the agile process herald the lack of documentation as a great benefit. We disagree with this view and note that the agile approach may actually generate more documentation than the waterfall model. For this, we note that there are several kinds of documentation. First, we must document the interaction of the users and implementors. This has legal and economic consequences as well a project management relevance. The agile framework generates significantly more documentation in this regard. Second, we must document the internal flow of the implementation inside the ticketing system so that the system functions. Particularly this includes how larger tasks have been broken into smaller chunks and how the emerging system is built from its chunks. This is an evolving documentation that is made larger through the kanban framework. Third, there is a functional documentation intended for the maintainer of the project on the implementors' side and the actual user in the users' organization. This is typically written only after the project is complete and is such much smaller than in the waterfall framework where it is written incrementally and then needs to be altered.

The agile/scrum/kanban combination almost always yields a result that solves the problem and makes all stakeholders happy with the result. It generally uses fewer economic resources and get the results done faster. The waterfall framework not only takes longer and costs more resources but it also usually generates bad emotions and conflict among all concerned.

We advise to use the agile framework and note in closing that this approach makes it impractical to agree *a priori* to a definitive deadline and budget. If these must be adopted, then the approach can still be used but there may come a time at which tickets must be rejected because they would break these restrictions.

3.2 Design of Experiments

Optimization projects nearly always involve data experimentally obtained in practice. Industrial practice does not lend itself to experimentation easily as this involves a large number of people and substantial cost. In industrial reality, making experiments can be very expensive due to lost production volume or quality if one intentionally puts the plant into a poor state. Thus the number and length of experiments should be minimal. In many cases, historical data is sufficient and leads to the complete absence of experimentation for the purpose of the project. The theory of *design of experiments* exists to minimize the number and size of experiments relative to a certain expected outcome of the experiments [75].

A good experiment will have the following features from the *scientific method* [105]

1. Reproducibility: The experiments are conducted *and documented* such that they can be reproduced by someone else at another location, at a different time with a different apparatus and this replicator will obtain similar results.
2. Comparability: The results are obtained in such a manner that they may be compared against other datasets collected by others previously or later. For this reason, measurements must have understandable units and any measurement or analysis process must be sufficiently documented to allow direct comparison.
3. Replication: A single measurement should be taken several times both to reduce and to measure the uncertainty with which the measurement process itself acts upon the data. For instance, we are to measure the length of a book with a ruler that has one notch every centimeter and no others. Sometimes we will measure 25 and sometimes 26 centimeters if we repeat the process. This is natural because we have no finer markings. Thus, the measurement uncertainty is ± 1 centimeter. Repeating this often will yield a measurement distribution that can be very educational.
4. Representativeness: The data collected must be representative of the problem under study. Please note that if a problem occurs with one car out of 100 cars, then taking 1000 random cars will not produce data representative of the problem; it will produce data representative of the situation at large. To obtain data representative of the problem, you should select 500 random troubled cars and 500 random untroubled cars.
5. Unbiased: The experiment itself must not pre-suppose a certain desired result or intentionally direct the measurement process towards a certain result. This is most typical in questionnaires to human responders. If we ask "you do agree, don't you?" then this is a leading question in that we predispose the responder to say yes.
6. Controlled: Any conditions that are not under study must be controlled for the experiment, i.e. we must strive to keep them constant while the experiment lasts.

However, a good experiment is not a designed experiment. To qualify for a designed experiment, it has to satisfy additional features,

1. Grouping: The individual measurements to be taken must be divided up into groups such that all the member of each group are comparable, i.e. they do not differ in any important way from each other. This leads to the assumption that the members of each group are statistically distributed in the same way.
2. Orthogonality: When changing one independent variable, this change should not have any effect upon the other independent variables. This leads to the possibility that the groups are independent of each other.
3. Factorization: Each independent variable is assigned several "levels" for investigation. Then each combination of every level of every independent variable is made into a group and many measurements for each group (presumably the same number for each group) are taken. For example, if we have three variables and two levels per variable, then we have 8 different groups.

The two first factors in a designed experiment mean that the groups of measurements are *independent and identically distributed*. This is an important statement in statistics. You will usually find it abbreviated as *iid* in books because it occurs so often. It means that each measurement is independent of another and that all the members in any one group are distributed the same (usually normally).

The third factor allows us to determine the influence of several independent variables upon an effect and upon each other (compare with the assumption of orthogonality) using the same number of trials that were necessary to study the effect of a single variable upon the effect.

The approach of design of experiments is very simple. The actual practice of it is complex because of having to make sure that all of these features actually hold in the practical application. We urge you to plan an industrial experiment well and prepare to test these features on test datasets acquired under normal production situations before going into the real experiment. Much wailing and gnashing of teeth may be thus avoided.

3.3 Prioritizing Goals

When facing the task to prioritize a list of objectives, we frequently do so emotionally and assign the highest priority to the objective that we "feel" is most important. This can lead to a misrepresentation if our opinion is wrong. Moreover, if the list of objectives includes 7 or more items, then human judgment has been proven to be suboptimal. The following method is designed to be easy to implement and will yield a much more objective result [107].

First, list all objectives clearly.

Second, compare every pair of objectives in turn. Note that if you have n objectives, there will be $n(n-1)/2$ such pairs. For every pair ask the following questions: (a) Which objective is more important? (b) Which objective is easier to attain? (c) Which objective will give the greater yield? We create three columns for each objective and put a mark into the importance column for the more important of the two objectives, a mark into the difficulty column for the easier of the two objectives and a mark in the yield column for the objective with the greater result. Note that since each objective is being compared to all others, any specific objective may collect quite a few marks. Next, we count the marks in all columns for each objective. The result of this might look like table 3.1.

Third, decide how relevant the importance, difficulty and yield levels are to you. We do this by assigning a weight to each of these three categories. The weights must add up to one. For instance, if difficulty and yield are equally relevant and importance is twice as relevant, then the weights are 0.5 for importance and 0.25 each for difficulty and yield. This decision is a subjective one based on the problem at hand.

Fourth, multiply each mark-count by its weight. For each objective, add up the three mark-counts that have been multiplied by their weight. This result is the priority score. The higher this score, the more relevant the objective to the problem. This has been applied in table 3.1 in the last column.

Fifth, this analysis not only yields an ordering on the list of objectives but a numerical measure of relevance. In our example in table 3.1 we discover that there are two equally relevant objectives (A and C), one objective that is just a little less relevant (B) and one objective that is much less relevant (D). This numerical score can be used to delete some objectives that obtain a very small score from further treatment.

Apart from the more objective nature of this process, it is the numerical score that is a major advantage over a subjectively sorted list. This approach can thus be used to shorten and rank a complex (even hierarchical) list of objectives.

Objective	Importance	Difficulty	Yield	Score
A	2	1		1.25
B	1	1	1	1.00
C	1	2	1	1.25
D			2	0.5

Table 3.1 The working table for an objective method to prioritize objectives. The leftmost column lists the objectives. The next three columns include the number of times that this objective was (a) more important, (b) less difficult or (c) more promising than another objective to which it was being compared. The last column contains the computed priority score.

Chapter 4
Pre-processing: Cleaning up Data

4.1 Dirty Data

The basis of any empirical research is formed by the raw data taken from the experiment. When we are dealing with laboratory experiments, we expend much effort in order to create an environment where few external conditions affect the data. In this way, whatever is seen can usually be assumed to be an effect of the experiment and not an effect of the environment.

In the industrial context however, the plant is usually built already and we must deal with whatever we find in place. Even if the plant is still in the design phase, production is the goal and the measurement process of data is a subsidiary goal. The environment interacts with the plant in myriad ways, e.g. raw material delivery, end product shipping or the weather. In addition to these meaningful interactions we obtain a number of unwanted effects. For example,

1. Sensors may stop sending data because they break.
2. Sensors may give out faulty signals because the sensor is malfunctioning.
3. Sensors may send wrong data because they have been clogged, gotten dirty or overheated.
4. Sensors may send unexpected data because they have been installed incorrectly in some manner, for instance in the wrong location (e.g. an air-temperature sensor placed in direct sunlight will yield data incomparable to normal air-temperature measurements that are taken in the shade).
5. Various temporal effects may lead to an occasional spike in value registered by the sensor that is however unphysical.
6. Data may be misinterpreted if it has been stored in the wrong units, e.g. pounds versus kilograms or Newtons versus kilo Newtons.
7. Various programming effects in the control and/or archive systems of an industrial plant may lead to values being distorted or stored in a way prone to misinterpretation.

P. Bangert (ed.), *Optimization for Industrial Problems*,
DOI 10.1007/978-3-642-24974-7_4, © Springer-Verlag Berlin Heidelberg 2012

8. The entire measurement process from sensor to archive system contains many sources of noise that make a signal look significantly more variational than it fundamentally need be.

The raw data must be cleaned before analysis so that we may draw conclusions that apply to the plant and not to the various sources of errors that will exist in real life. This chapter discusses some methods used to clean data.

We distinguish several fundamentally different ways to prepare or pre-process data for subsequent computational methods,

1. Discretization: These methods convert a continuous stream of data into a sequence of data points and so select when a new data point is added to the sequence.
2. Outlier detection: These methods exclude unwanted data points from further processing.
3. Data reduction / Feature Selection: These methods group similar data points into clusters to reduce the total volume of the data.
4. Smoothing: These methods change the measured data points in order to remove noise and other atypical components.
5. Sampling: These methods choose the data points presented to the computational method so as to represent those types of events that are most relevant to the intended goal of the analysis.
6. Interpolation: These methods supply a value for a variable in between several measured values.

4.2 Discretization

4.2.1 Time-Series from Instrumentation

Many industrial sensors are analog sensors in that they produce a value in the form of an electrical signal at the rate of 50 or 60 Hertz. With many sensors, e.g. temperatures, the vast majority of signals are identical and so there is no need to store all of these values. With other sensors, e.g. vibration, this rate may be sensible for certain types of analysis. In any case, we cannot usually store data at this rate and do something useful with it.

The analog data must be digitized and as such we must answer the question of when we should accept a measurement as sufficiently new to record. There are three basic methods that are common in the industry.

First, a value is recorded every x seconds. The size of x is chosen manually at installation time. This has the advantage that data is equally spaced in time, which is good for many machine learning algorithms. It has the disadvantage that it may easily disregard important events or create much unnecessary duplication if things do not change for some time.

Second, a value is recorded if it differs from the previously recorded value by either a difference x or by a ratio y. This is the approach taken by most archive systems that make the interpretation that a value does not change until a new value is recorded and thus, it is claimed, the system records everything while minimizing disk space usage. Indeed, this is a good approach if x or y are chosen appropriately. Sometimes they are chosen without good reason or for storage reasons and not based on the equipment. The setting of x and y should be a result of the measurement accuracy of the sensor for the process in which it is installed. Only then, can we say that we actually record all the dynamics.

Third, we record a new value every x seconds unless the value is outside certain boundaries in which case, we record a new value at a different (usually faster) rate for a given time interval. This is often done by control systems to create a kind of snapshot of atypical conditions. The reason that this is done is often for regulatory reasons in that the operator must keep detailed records of potentially dangerous events. Such snapshots do not usually have much scientific value for modelers but they do offer value to engineers constructing physical solutions.

From the point of view of machine learning, the discretization should record all important features in the data stream and, if possible, not store too many instances of the same feature. Based on this loose criterion the second method above is a good method. This method still stores a great many instances of the "normal" operating condition of the system and thus duplicates this feature but it only stores it if these instances are interrupted by some other feature. As we are dealing with time series, this is very relevant and should thus be kept.

4.2.2 Data not Ordered in Time

In the industry setting, virtually all measurements are time-series in that a sensor produces a data stream of values that are sensibly ordered and indexed by the time at which the reading was taken. Data that is fundamentally not ordered in time, such as addresses or names, do not generally occur in the industrial setting and will be ignored in this book. Even the retail purchases of a customer to be treated later in this book are events ordered in time that indicate time-dependent behavior and interest dynamics.

Supposing we disregard the time order in the above data and simply consider the data as different measurements taken at the same point. We may now discretize this data further. The most popular method is called equal-size binning. We select the maximum and minimum value of the feature over the whole data set and break this interval into a chosen number of segments. Every observation falls into one of these segments. We can now count how many observations occurred per segment and divide this count by the total number of observations. This procedure yields a probability distribution function (pdf) over the possible feature values. Such a pdf can yield useful information for the human analyst. For example one may diagnose the type of distribution function, e.g. a normal distribution, and draw conclusions about the noise source on this basis. This pdf can also yield useful information for an automated system if we continuously compute the pdf and monitor how it changes over time; see chapter 5 for details.

Often it is hard to tell the difference between different pdf's based on the empirical distribution computed as above because the distributions differ mainly in the tails and these are not represented at a high resolution with the method of equally spaced bins. There are a variety of methods that adaptively change the size of the bins in such a way as to achieve a very representative picture of the distribution. A detailed presentation of these would lead beyond the scope of this book and we refer to the literature for further details [60, 93, 48].

4.3 Outlier Detection

We want to smooth the data incoming by deleting from it extreme values, extreme changes and so on so that we can delete the "noise" that is introduced by various factors other than the effect we wish to study. This cleansing is crucial for the noise would overshadow any analysis or conclusion and possibly lead to errors. In figure 4.1 we present an example with the original data in gray and the denoised data in black dots over it. You can see that the underlying dynamics is preserved but the majority of the fluctuations have been removed.

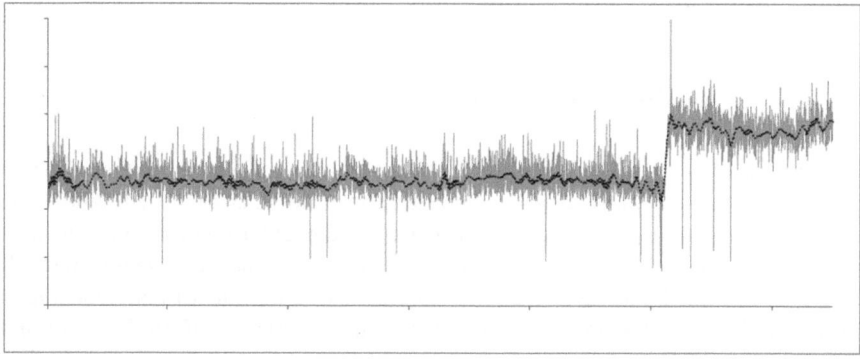

Fig. 4.1 This dataset is a vibration measured on an industrial crane during its operations. The sharp increase in vibration represents an acceleration. The raw data in gray indicates a significant source of noise versus the cleaned black curve.

There are various types of outliers that will be treated in the sections below.

4.3.1 Unrealistic Data

A reactor temperature in a chemical process of 20 000 degrees Celsius is dangerous. Indeed it is unrealistically high. Perhaps this is the result of a floating point to integer conversion or some other data loss along the way but somehow two extra zeros got past the system. This kind of value must be extracted as unrealistic. If an automated system were to accept such a measurement as a true result of a faithful measurement process, it would lead to an automatic shut-down of the plant and thus cause significant damage.

Percentages below 0 or above 100 are not necessarily wrong but they are suspicious as this is not the usual intent of percentages. Temperatures below 0 whose unit is in degrees Kelvin pose a problem. Either the units are wrong or the values are wrong.

In order to filter out unrealistic data, we introduce a minimum and maximum allowed value for every sensor. Any value outside this range is automatically excluded from further processing.

4.3.2 Unlikely Data

If we introduce a probability distribution for each time-series as discussed above in section 4.2.2, then we can assess each value in terms of the number of standard deviations that it is away from the mean. We may then set limits on this distance from the mean.

If the distribution is normal, then most values should lie within a few standard deviations of the mean. Thus, if we draw the line at, say, three standard deviations, then we should be excluding only a small fraction of all data points. This does amount to a strict upper and lower limit on the value just like in the previous section. There is one major difference however: As the mean and standard deviation change over time, so does the upper and lower limit to reflect that change. This approach yields an adaptive hard border.

Please note that many probability distributions are not symmetrical about the mean and thus the limit (expressed in multiples of the standard deviation) must be specified for both directions away from the mean and will in general have different values so that we do not over-represent one side of the mean.

We use these methods to remove large spikes from the data that are very unlikely to happen and generally represent sensor glitches.

4.3.3 Irregular and Abnormal Data

Using a clustering method, we may construct clusters of data points. Most data points should then belong to a cluster together with many other data points. Each cluster represents a particular operational condition. Some data points will lie outside any cluster and thus represent solitary operational conditions. These data points are, by definition, irregular as they do not fit into any of the regularities that are represented by the clusters. Such points are then excluded from processing.

There are many such clustering methods. We will introduce k-means, the most popular clustering method in practice, in section 5.3.4.

Using various statistical methods, we can introduce maximum and minimum limits based on more intricate measures than mentioned above. We treat three such methods in sections 4.5.2, 5.3.5 and 5.3.6. All of these methods transform the actual measurement into a statistical quantity over a certain time windows, which induces a new time-series. This is now given a minimum and maximum limit and the data points associated with the out-of-bounds statistical quantity are excluded.

The methods of this section are relevant if we want to exclude data that comes from an operational state that represents a working process in a state that is atypical or considered bad. We restrict our attention therefore only to the really normal and wanted operational mode.

4.3.4 Missing Data

Values that are missing pose a significant problem to the data analyst. Various of the previously discussed problems are effectively missing data. For example when a sensor malfunctions and gives unphysical data, then this is effectively a missing data point. We also have data that is genuinely missing, for example when sensors are fully broken and do not output anything.

We must decide whether it is feasible to replace the missing values by some sort of computed values. Frequently this is possible by interpolation (see section 4.7) but sometimes we must get more intelligent and actually use a model of the missing object with respect to the data that we do have in order to plausibly fill the gap. This method, if it is truly reliable, may in effect amount to a soft sensor and thus this variable may not be needed at all (see section 4.4).

Most of the time, however, the replacement of the missing data points with sensible data points is something that must be decided upon the merits of the individual problem at hand. If it concerns a time-series and the gap is short, the solution is usually interpolation. If the gap is long, then it may not be possible to fill without a reliable soft sensor. Often times, the pragmatic solution is to exclude these data points from analysis altogether. However, this may introduce a bias into the analysis if there is some form of structured cause for the absence of the data.

4.4 Data reduction / Feature Selection

4.4.1 Similar Data

In a normal industrial setting the vast majority of data points will be very similar. Indeed, with respect to sensor uncertainties and drifts, the vast majority of data points will even be identical. Billions of data points may, in fact, represent only hundreds of truly different conditions. The dynamics of these hundreds of conditions may be suitably represented using thousands of points. Thus, we could, in principle, reduce the volume of billions of points to millions of points without actually loosing any information.

Please note here the difference in meaning of the terms *data* (which we loose in this operation), *information* (which we retain) and *knowledge* (which we hope to generate using methods of other chapters).

Such a dramatic reduction in data volume has several advantages. First, many machine learning methods are so resource intensive that they become practically unusable over a certain data volume. Second, machine learning takes an amount of time that is a (not necessarily linear) function of the input data volume and so learning speeds up. Third, the computational resources (computers, chips, hard drives, electricity etc.) that must be expended to produce and maintain the model reduce.

A simple method is to introduce a distance measure between data points. Then we compute the distance between a data point and the others. If the distance between it and some other point is less than a certain limit value, then this data point is deleted because it is sufficiently similar to an existing point. For statistical information, we may introduce a count for each remaining data point that shows how many fellows were deleted when encountering this point. This count may then be used to give this point a statistical weight with respect to the method below.

This method is simple but it is expensive as the number of distance function evaluations is quadratic in the number of data points and if we have billions of them, this may take a long time in practice[1].

In the context of time-series however, this is a tricky issue as deletion of data results in a loss of information as a repetition of a measurement is not actually a duplication due to the fact that the second measurement occurs at a different time. Deletion would thus disturb the flow of information about the evolution of the system. We must thus be careful if the data points are not independent of each other.

4.4.2 Irrelevant Data

In addition to data points that are so similar to others that they can be ignored, we have the possibility that data is irrelevant. Irrelevancy is a property not of a data

---

[1] Please note carefully that when we speak of long computation times here we do not mean the annoying minute that a computer takes to boot but rather computations lasting many days or weeks.

point but of a variable. If a variable has no influence on the output of the model that
we desire to create, then this variable will not be required for training and is thus
irrelevant. Then we may exclude this variable altogether – from every data point.
This not only reduces the data volume but it reduces the dimensionality of the data
set, which is a very valuable feature.

This may be determined by computing correlation coefficients and excluding
variables with a coefficient lower than a cut-off.

4.4.3 Redundant Data

Variables are redundant if they can easily be inferred from others. A frequent observation in the industry is that important temperatures are measured with three separate sensors. The reason is that if one sensor breaks down, the other two still deliver
good values. Thus, the temperature is assumed to be equal to that measurement that
is delivered by at least two agreeing sensors. For our modeling purpose, it would be
a waste to have three temperature variables as the combined temperature signal will
cover the situation. Having three sensors solves a physical problem (sensor failure)
but is not useful for data analysis and so two variables are redundant.

Some chemical processes laboriously take samples of the substance and have it
analyzed in the laboratory. In most situations, it is possible to model the laboratory
result using process measurements and to introduce what is called a *soft sensor*. This
is a software that computes the outcome of what would have been a physical measurement. Based on various means such as past data and knowledge, it is possible
to craft a computational method that makes the physical measurement superfluous
and we have the outcome of the computation only. As this proves that this variable
can be derived from the others, it is redundant for modeling purposes. This does not
reduce its value for the human operator who may very well be interested in it but we
do not need it for modeling and thus reduce our dimensionality.

4.4.4 Distinguishing Features

A *feature* is a macroscopically different and interesting state as opposed to other
features. We might wish to distinguish normal operations from abnormal operations,
good batches from bad batches and the like. Sometimes we have only two features
(good vs. bad) and sometimes quite a few (abnormal for this and that reason). We
are looking to distinguish these states from each other using process variables, so
that we can detect such states automatically, and perhaps even ahead of time.

For the purposes of this section, we will assume that we have two features. Figure
4.2 displays several possible methods that one might use to effect a separation. We
instantly observe that certain methods apply to certain situations. In this case, the
data is the same in all images and we see that (a) and (c) do not effect a perfect sep-

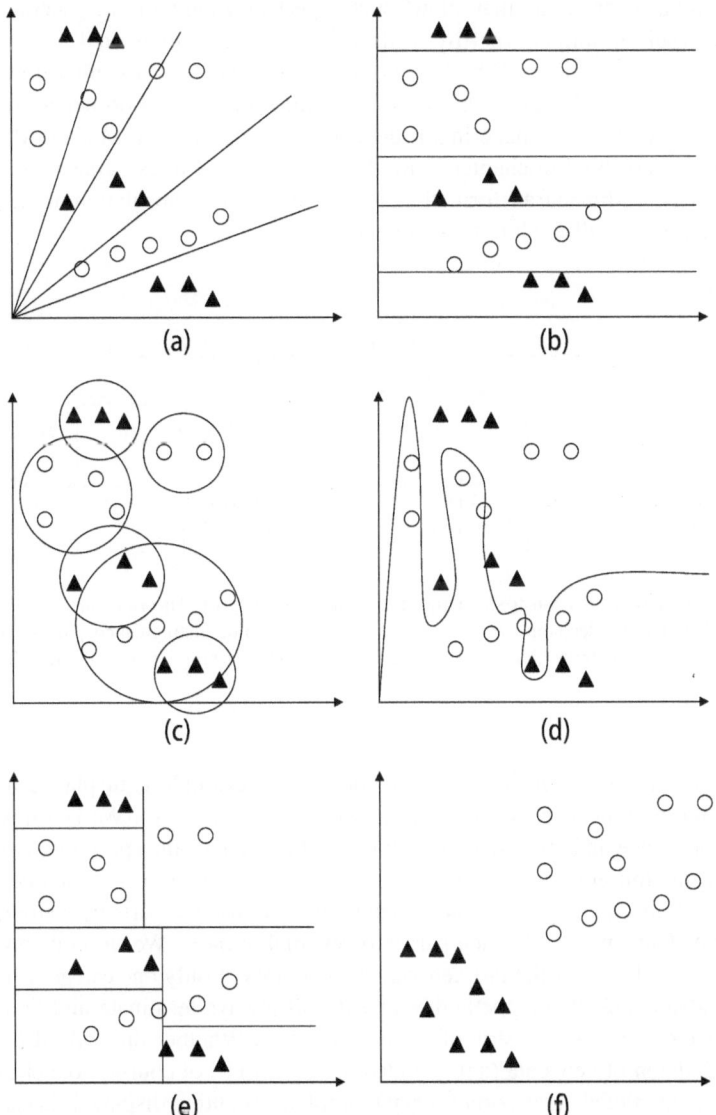

Fig. 4.2 Some different possibilities to distinguish between two categories: (a) Lines emanating from the origin, (b) lines parallel to the axis, (c) clustering methods, (d) curves, e.g. neural networks, (e) decision trees and (f) an ideal arrangement achieved for example by a coordinate transformation.

aration. The method (b) is clearly simplest, method (e) also quite simple but method (f) is the most impressive as it requires no more computing once a coordinate transformation has been found that allows such a perfect separation. In general method (f) is the result of human expertise while all others can be automated.

It is the responsibility of the human miner to decide which method to apply in the particular case. Methods (a) and (b) are quite simple and do not need further discussion. Method (c), clustering, is described in section 5.3.4. Method (d), neural network, is described in chapter 6. Method (e), decision trees is described below. Method (f), coordinate transformation, is effectively an encapsulation of the experts expertise and so is difficult to treat in generality.

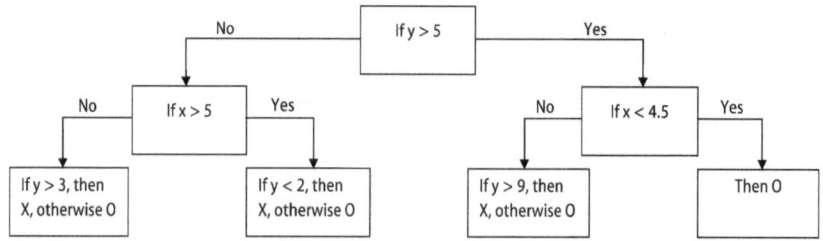

Fig. 4.3 The actual decision tree for the image in figure 4.2 (e). The rules are easy to follow and lead to a definite decision regarding the two categories to be distinguished. The values in the decision rules and the branching can be learned by a machine learning algorithm trained by sample data.

A decision tree is a relatively simple object. An example is displayed in figure 4.3 and the graphical result of it can be seen in figure 4.2 (e). We begin with the entire state space and introduce a dividing line that divides the space into two parts. Whether this initial division is a horizontal line or a vertical line is the decision of the miner. Subsequently, each part is again divided into two parts by a straight line and so on. Thus, we divide each part into yet smaller parts. We stop this process if and only if each part of the last generation has points of only one category in it.

The advantages of this method is that it will always terminate and that it will always yield a perfect division of the training data. Whether this will also yield a perfect division of new data (not included in training) is, of course, not certain.

Once constructed, a decision is easy to implement and to display. It is easy for a person to follow and to understand. In many cases, it may even be constructed by hand.

4.5 Smoothing and De-noising

4.5.1 Noise

Noise is any unwanted perturbation in a wanted signal. The classic hissing static in a poorly tuned radio is an example. It is difficult to hear the radio program because the noise disturbs it. This can occur in any data recording. In figure 4.4 there is an example of the noisy signal on top and the pure signal on the bottom. We note that the noisy signal seems to have no structure but the pure signal does. Thus, in this example, the noise is so dominant that the signal is almost completely swallowed.

Noise occurs in the industrial context just the same. There are indeed many sources of noise from microscopic variations in the actual physical medium to be measured, over electronic transmission all the way to the conversion of a 50 or 60 Hz analog signal to digital data. We may get spikes in the data or see a drift that are due to measurement effect rather than a true process cause.

For analysis, we must divide signal from noise. A method that can do this is called *de-noising*. In pictorial terms, the task is to convert the top part of figure 4.4 into the bottom part. This should be a conversion of data into different data while actually retaining the information content. This is not strictly possible, as we do not know the exact nature of the noise. The noise must therefore be modeled or guessed at in some algorithmic fashion in order to be able to subtract it. The field of electronic engineering has many methods to deal with this that are grouped under the heading *filters*, see [120] for more details. Such filters analyze the noisy signal and only let portions of the signal through while removing the remainder. It is beyond the scope of this book to discuss filtering theory. In general, using filters requires at least some knowledge about the noise as the filter must be tuned.

A very popular method of de-noising is the *moving average*. We replace each point by the average of the last x points. This is used almost exclusively in the stock market and similar financial analysis. This is by far the simplest method and can be easily done on a spreadsheet but we must be careful with this. Most naturally occurring data sources (including the people that make up stock market prices) act on several scales. There are features that are short-lived (e.g. they go from cause to effect in a day) and some are long-lived (e.g. they go from cause to effect in weeks). If you use a moving average, the parameter x defines your analysis scale. You will be deleting all features with scale less than x and you may easily misrepresent features with larger scales. Thus, moving average is a suitable method only if there is only one scale present (and this is rare) and you know what it is (and this is even rarer).

In the next section, we introduce a very general method of denoising that can be applied to any data and that is quite powerful in absence of knowledge about the sources and nature of the noise present. If such knowledge exists, we recommend an expertly designed filter. Other ways of denoising include all the methods of interpolation discussed in section 4.7.

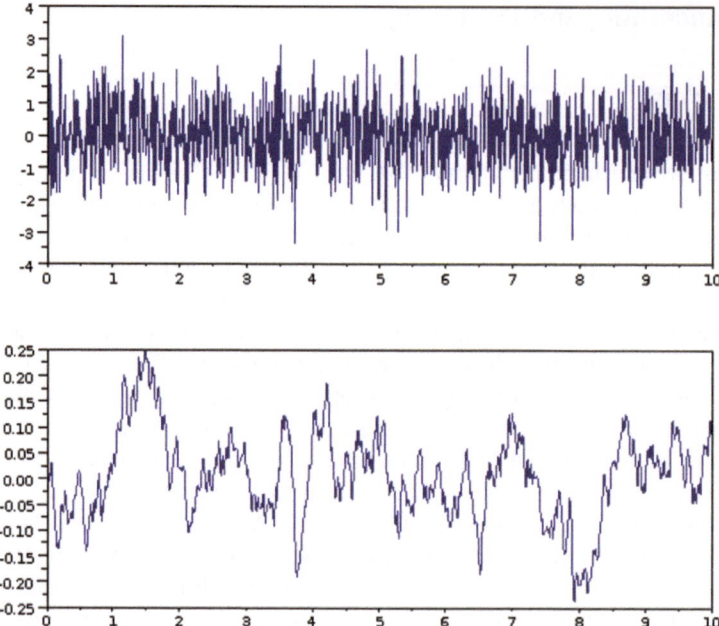

Fig. 4.4 The noisy signal on top and the pure signal on the bottom.

4.5.2 Singular Spectrum Analysis

The method of *singular spectrum analysis* (SSA) embeds the data in a higher dimensional space with the intent of then searching for a new coordinate system - a set of orthogonal vectors to form a basis for the larger space. This basis is chosen according to a number of mathematical criteria to isolate the principal directions that are important, i.e. common, for the signal observed in the past. Having performed this analysis, the method computes the eigenvalues of the new basis matrix. These eigenvalues are effectively the directional variances of the signal along these newly created basis vectors.

As the largest eigenvalue typically contributes more than 90% of the signal strength, it is reasonable to focus only on this value for the purpose of identifying the signal abnormality. The value is therefore the largest directional variance observed along the principal directions and is associated with a particular direction, i.e. a particular weighted linear combination of the original time-series.

If the variance is low, this means that the signal (along this direction) is tightly clustered around a particular numerical value. If the variance is high, then this same signal varies more widely. A temporary jump from low to high variance thus indicates the presence of outliers, which signal a form of abnormality. A temporary change from high to low variance indicates that the system found a stable configuration. Whether this stability is good or bad for the system cannot be interpreted here and so it will be flagged as a visible change. A more permanent jump from low to high variance, and vice-versa, indicates a structured qualitative change in the system state. The directional variance computed forms itself a time-series with a mean and standard deviation. We can use SSA to de-noise a dataset most simply by excluding the original data points that, in the SSA analysis, yield an eigenvalue larger than several standard deviation away from the mean.

Once the signal is broken into its various components, the signal can be reconstructed leaving out the noisy parts. This is the way that SSA can be used for smoothing operations. Due to the highly sophisticated analysis that SSA performs, the method can also be used to other purposes such as time series analysis (e.g. splitting into seasonal and global trends) or even prediction.

Figure 4.5 displays an example of the variance change for a particular dataset. We note that the principal variance exceeded the threshold at time-window 22 and then settled back down again. This event would be flagged as abnormal and thus removed during pre-processing.

We note in passing that the coordinate system referred to is determined afresh at each time-window and so changes its identity as well. A more thorough analysis is possible in which the temporal evolution of the coordinate system is analyzed alongside the variance along its major component.

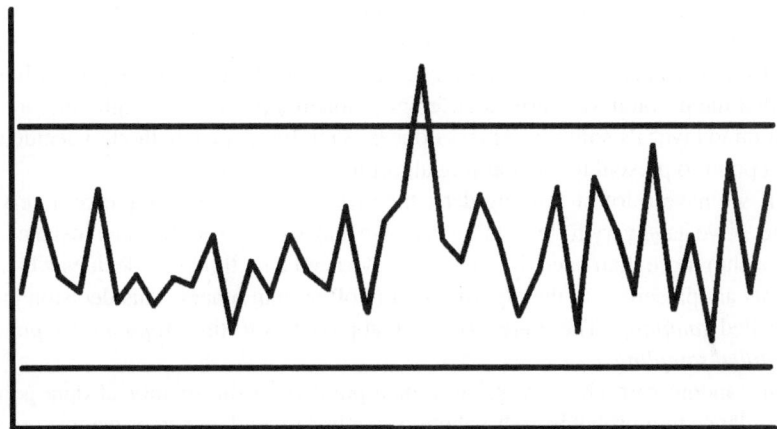

Fig. 4.5 Development of the abnormality score (SSA) for a single sensor for a consecutive sequence of time windows.

The exact details of the method are too mathematically complex to be described in detail here. The interested reader is referred to the literature [67].

4.6 Representation and Sampling

In order to make any useful data analysis possible, the data must be representative of the problem. Take for instance the problem of detecting scrap parts in a manufacturing process. If the process produces approximately one percent scrap, we can easily construct an algorithm that is right 99 percent of the time: The algorithm always says that the part is good. This method will, of course, not find any scrap parts but as these only make up one percent, the method is right 99 percent of the time. Therefore, the measurement of how many parts the method identifies correctly is the wrong measurement in this case. Training an algorithm with an overwhelming number of samples of one kind and very few of another kind is the wrong approach to take.

We must collect data in which the problem is clearly visible. The problem of representation is different than that of noise discussed above. Noise is the intrusion of an unwanted external effect to mess up the data, e.g. the hissing noise muffling a radio signal. Problems of representation come from within the process and contain actually meaningful data for the process – just not meaningful for the problem under investigation.

We solve the problem of representation in two mutually supporting ways. First, we only record data that is useful for solving the problem. The decision of how exactly to do this must be taken dependent upon the problem and cannot be discussed in this book. It is however essential that a expert with experience carefully designs a measurement method for this purpose. Second, the obtained data is carefully vetted so that the problem (or various different problems) receives the right amount of attention to favor its solution. This is done by sampling, i.e. a method to decide which data point to present to the learning algorithm at any time.

If we have a lot of training data, the process of training may take a very long time. If we have very little training data, we may have to present each instance to the algorithm more than once. In both cases, we need a method that decides which data points are presented to the algorithm more often than others. This decision process is called *sampling*. There are two basic approaches to this: *random sampling* and *stratified sampling*.

In random sampling, we select a data point from the totality of data points in accordance to a probability distribution. We must decide whether we want to allow a data point to be selected more than once (sampling with replacement) or to be selected at most once (sampling without replacement). We must also decide what the probability distribution is going to be. Generally, one takes the uniform distribution (all options are equally likely) and samples with replacement; this is also the simplest of all sampling approaches to implement.

We choose random sampling when we have a huge data set and must train on a subset to save training time. It is assumed in this approach that the data is sufficiently representative of the problem we wish to solve. If the data is not representative, then we use the second sampling approach.

In stratified sampling, we select a data point in much the same way except that the probability distribution is chosen to over-represent rare data points. In industrial systems there are usually rare circumstances that are quite essential to be understood and modeled, e.g. outages or shutdowns. If our data set contains a large proportion of very similar data that represents normal conditions and a small proportion of diverse data that represents abnormal (but very interesting) conditions, then we should use stratified sampling. This would then select the abnormal points with much higher likelihood than the normal points.

The definition of "normal" and "abnormal" is, of course, essential for stratified sampling and usually we choose a numerical measure that provides a sliding scale as opposed to a binary division. The methods of section 4.3 above provide suitable measures. We may also cluster the data into various operational conditions and then select an equal number of points from each condition regardless of how many distinct points are actually in each cluster.

We choose stratified sampling when we have situations that are important for modeling that are rare in the data set. This is the general situation in the industry and so this is the method of choice for this book. A statistical measure supplies an abnormality index of a data point and this index guides the likelihood with which it is presented to the training algorithm. Exactly how this scales is a matter of some care and some balance as to how important this abnormal state is in comparison to the normal state as we do wish to represent both. This balancing act is best carried out by an experienced analyst.

4.7 Interpolation

Suppose we have taken several measurements of a dependent variable at regular intervals of the independent variable. Now we ask for the dependent variable at some value of the independent variable that lies in between two previously measured values. As the point in question is within the range explored already, the answer to this question is known as *interpolation*. If the point in question were outside the range explore, then we would refer to it as *extrapolation*.

Interpolation and Extrapolation are very different. In general interpolation is much simpler as we can rest our computations upon firm evidence on either side of the value demanded while in extrapolation anything could happen outside of our experience.

The general approach is always the same. We choose some function that we believe has the capability of representing the actual source of the data and adjust the parameters of this function until it fits the data obtained. Then we use the function to compute the desired value. In figure 4.6, we see an example. The black circles are the actual data. Both curves are attempts at fitting a function to the data. Please note that both curves fit the data perfectly. Yet they differ significantly in terms of what answers they would give to the interpolation question (at least far away from the central portion). We do not have any way, based on the data alone, to distinguish between the two functions, i.e. to call one "better" than the other. We may use Occam's razor here to choose the simpler of the two on grounds that if in doubt, the simplest explanation should always be chosen, but this is no guarantee.

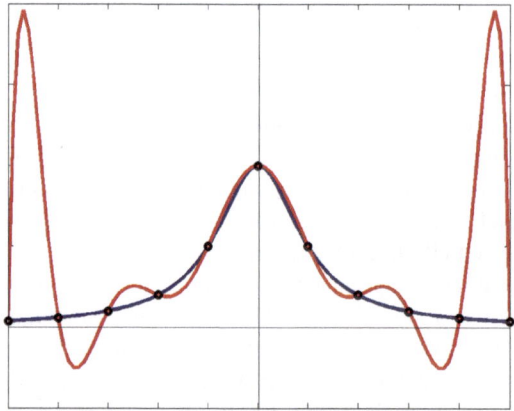

Fig. 4.6 The black points have been measured and two curves have been attempted for an interpolation. As both curves fit the data perfectly, we cannot choose between them except by adding an external criterion or knowledge.

Methods of interpolation thus come down to choosing a function believed to be meaningful for the data. We discuss some popular choices below. We will assume that we have measured points $(x_1, y_1), (x_2, y_2), \cdots (x_N, y_N)$ and wish to find a function that best represents these measurements.

The *polynomial* looks like this

$$P(x) = \frac{(x-x_2)(x-x_3)\cdots(x-x_N)}{(x_1-x_2)(x_1-x_3)\cdots(x_1-x_N)} y_1 + \frac{(x-x_1)(x-x_3)\cdots(x-x_N)}{(x_2-x_1)(x_2-x_3)\cdots(x_2-x_N)} y_2 +$$
$$\cdots + \frac{(x-x_1)(x-x_2)\cdots(x-x_{N-1})}{(x_N-x_1)(x_N-x_2)\cdots(x_N-x_{N-1})} y_N$$

(4.1)

according to Lagrange's classic formula. This presentation fits all the points exactly and the polynomial will have degree $N-1$.

A *Fourier series* looks like this

$$f(x) = \sum_{n=-\infty}^{\infty} c_n e^{inx} \quad \text{where } c_n = \frac{1}{2\pi} \int_{-\pi}^{\pi} f(x)e^{-inx}dx$$

(4.2)

that is generally capable of representing any periodic phenomenon.

Both the polynomial and the Fourier series are functions that are meant to be applied to the whole data set. The reason is that if we were to apply them to parts of the data set, then the functions of the different parts would not agree on the boundary and we would have a non-continuous function. This is generally not desirable as nothing physical is really non-continuous.

So if want a piecewise defined function, we must have a method for making the pieces agree on the boundary. This is the job of *splines*. Most popular are the quadratic or cubic splines, which are simply quadratic or cubic polynomials valid for a part of the data. We select a few points, fit the spline to it subject to the condition that the value on the border agrees with the value of the next spline. Thus we simultaneously fit a number of polynomials to different parts of the data and thus produce a continuous piecewise defined function. An example looks like:

$$S(x) = \begin{cases} -2 - 2x^2 & 0 \le x < 1, \\ 1 - 6t + t^2 & 1 \le x < 2, \\ -1 + t - 2t^2 & 2 \le x < 3. \end{cases} \tag{4.3}$$

For most applications in practical industrial computing, the cubic spline is the interpolation function of choice. If domain knowledge is available and the function that the effect should have is known, then we can use the theoretical function, of course. Generally this is not known however and we must work with a pragmatic choice. The cubic spline is a pragmatic choice because it is quite robust and practical to evaluate apart from its desirable properties such as continuity.

4.8 Case Study: Self-Benchmarking in Maintenance of a Chemical Plant

Co-Authors: Claus Borgböhmer, Sasol Solvents Germany GmbH
Markus Ahorner, algorithmica technologies GmbH

We propose that popular benchmarking studies can be augmented or replaced by a method we call self-benchmarking. This compares the present state of an industrial plant to a state in the past of the same plant. In this way, the two states are known to be comparable and it is possible to correctly interpret any changes that have taken place. The approach is based on data-mining and thus can be run regularly in an automated fashion. This makes it much faster, cheaper and meaningful than regular benchmarking. We demonstrate on the example of maintenance in the chemical industry that this approach can yield very useful and practical results for the plant.

4.8.1 Benchmarking

Several companies offer benchmarking studies to industrial facility managers. These studies provide each facility with a questionnaire to be answered by the facility and then the benchmarking company produces a report based on the comparison of these questionnaires with similar facilities. Each facility can recognize its own position in the statistics as well as the values for the best-in-class and some other quartiles. Thus, the facility can see how it compares to similar facilities worldwide. Because this is done in many categories, one can deduce specific improvement areas for each specific facility and thus benchmarking is said to help each participant of these studies to improve.

While the idea of these studies is, as described above, quite sound, these studies suffer from several endemic faults. First, no participant knows the identity of the other participants and thus the definition of "similar" facility is entirely in the hands of the organizer. Second, as almost no facilities are truly comparable, differences in any one detailed category may be due to causes that are outside of the control of any operator and maintenance organization, e.g. engineering differences in the initial building phase of the facility. Indeed, the benchmarking study only allows conclusions about the nature of differences but not the causes (and thus remedies) for such differences. Third, many questions are not as precisely defined as they would need to be for true comparison. For example the question for the current financial value of the facility could be answered by (a) the known historical building cost, (b) the estimated current re-building cost, (c) the known current book value of the facility, (d) the estimated market value of the facility, and several more possibilities. Clearly, these numbers are quite different and thus lead to different conclusions in the study.

In conclusion, such studies produce results that may indeed lead to favorable changes but they cannot be taken at face value and indeed lack many of the truly interesting facts.

4.8.2 Self-Benchmarking

As an alternative or addition to traditional benchmarking studies, we suggest a method that we shall refer to as self-benchmarking, which simply refers to comparing one's own facility to itself at previous historical times and to itself at the current time in order to find developments and differences. We will demonstrate below how this leads to useful results.

It is apparent right away that this solves all four of the above problems of benchmarking studies: First, we know exactly who is taking part, i.e. all players in our own facility. Second, we are aware how comparable our own facility parts are to each other or how our facility compares to itself at an earlier time. Third, as the numbers are centrally collected by one party in the facility there will be no uncertainty as to the meaning and comparability of the numbers.

We note in passing that such a survey is also much faster and cheaper than participating in an external benchmarking study. If the data is set up appropriately within the ERP system of the facility, the study can be largely automated and repeated at monthly intervals to yield a running account of improvements and deficiencies. This is in contrast to external benchmark studies that consume much time and are done at much longer intervals in time.

We will illustrate the concept of self-benchmarking by using the maintenance department of the two sites in Moers and Herne of a chemical production company in Germany. Here, the process includes the following phases:

1. Extraction of all maintenance requests (German: Meldung), work orders (German: Auftrag) and the maintenance reports that allow the cost to be divided into human resource costs and material cost and further into internal and external costs.
2. Raw processing of this information so that requests, work orders and reports can be matched. It is useful to have a unique identification number in the ERP system for this purpose. Here the data is cleaned and prepared for further analysis by filling empty entries, deleting exceptional entries and performing a myriad of standard data cleaning and standardization procedures [107].
3. Aggregating the information along the following dimensions: by plant within the facility (one plant makes one product), by priority of the maintenance measure, by service types (e.g. repair, inspection, engineering), by request type (e.g. maintenance, shutdown), by planning groups, by divisions, by duration of the maintenance measures, by cost category (e.g. external material cost, internal human resources). In each dimension, we might tabulate the number of measures, the total cost and the average cost.
4. Step 3 can be done for equally sized historical periods so that a time trend is obtained and the facility can be compared to itself at an earlier time.
5. These aggregated data are then interpreted and combined into a report. The interpretation yields natural conclusions and suggestions for improvement, e.g. the relationship between reactive and preventative maintenance. If the raw data is clean enough and the categories are sufficiently detailed, then useful improvements can be directly read off this analysis.

This last conditional, of course, also provides suggestions for improving the data quality by introducing further categories or improving the data entry into the ERP system. In fact, we have often found that many users see an ERP system like a one-way street into which data disappears. From this point of view, it is not worthwhile to make a special effort to keep the data clean. With such approaches like we describe here, the ERP data becomes a two-way street by making such studies possible. This influences the ERP data entry and thus provides an improvement to the general data management of the facility.

We provide an example of this in figure 4.7. The vertical axis measures the relative number of work orders for each of the top three priority levels. The solid gray line is priority 3, the dashed line priority 2 and the dotted line priority 1. The horizontal axis denotes the different plants in the facility. The plants have been sorted in decreasing order of priority 3 work order weight. We easily see an inverse relationship between the priorities 2 and 3. We also see a tendency of priority 1 increasing as priority 3 decreases. We may definitely see a few plants whose priority 1 weight is far too high compared to the other two priority levels, given the general level. The

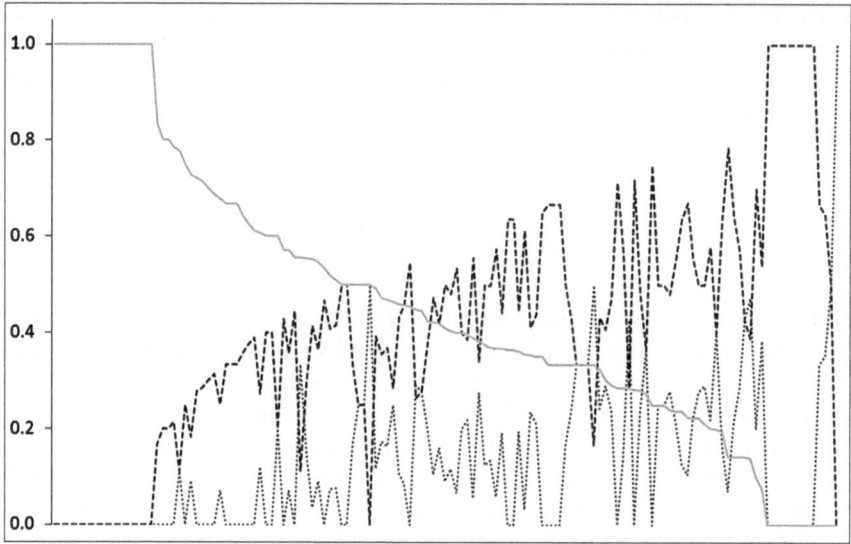

Fig. 4.7 We see the relative number of work orders of any one priority class relative to all work orders of a certain plant. Priority 1 is the dotted line, priority 2 is the dashed line and priority 3 is the gray solid line. The plants have been sorted in decreasing order of the priority 3 work orders.

average weight of priority 1 is 13% but there are plants with weight 50%. The higher the weight of priority 1, the less planning is being done and these work orders will cost more. Using this comparison, we can easily see which plants can be improved by comparing the facility to itself.

4.8.3 Results and Conclusions

Over the course of a year, the maintenance of a site implements many measures that yield a large enough pool of data to allow statistical conclusions to be representative and sensible. We must be careful to exclude outliers from such conclusions as they skew statistics into the meaningless. What an outlier is must finally be decided manually but can be pre-processed automatically. A very costly measure is a natural candidate for an outlier but it may not be if this measure is indeed regular.

After the data is cleaned such, we may compare averages by type with the average overall. In this way, we may discover that a particular equipment of the site is particularly cheap or expensive to maintain. A look into the measures will reveal what went on to yield this effect. Usually there are also particular types of measures that have caused unreasonably large costs. Thus, we can identify the so called "bad actors."

Often bad actors are those that failed often or those that necessitate particularly expensive measures. However, the bad actor really is that equipment that causes unexpectedly high total costs. If the costs are expected, it is not a bad actor but a normal actor. The costs can be accumulated over many faults or a few large ones. In total, this is something that must be extracted by an analysis.

While dividing the measures by priority, we note that the classification of a measure as first priority (to be done as soon as possible) or second priority (to start within three working days) is a significant element in predicting the cost of the measure. The same measure will be 43% more expensive if it is in first priority than in second priority. The reason for this is that the lower priority allows for planning and this reduces the amount of working time and sometimes also material used. We note at this point that it is possible to predict when machines fail a few days in advance, so that one could immediately transfer many current priority-one jobs into priority-two jobs and leverage this cost difference, see chapter 6.

From the analysis over types of jobs, one can see mainly the difference between preventative and reactive maintenance. The theory goes that more prevention will cause less reaction to be necessary. Clearly there is an optimum relationship between these two both from reliability and expense point of views. Taking historical data and making a sliding comparison in that history yields various observed relationships and will let the analyst conclude the true optimum point. This differs from plant to plant and thus can hardly be discovered in a benchmark study with different plants.

Something of interest to all facility managers is the question of whether the work is done by internal or external staff and whether material is provided by the stock or ordered in for the job. That is again a point in favor for self-benchmarking. It depends on such features as what the facility generally needs done as well as what the external but geographically local area has to offer in terms of service providers. We again recommend the self benchmark. Here we can see the evolution over time and find the optimum. If enough work exists, then it is cheaper to do this work with internal staff and if the failure will lead to relatively high production losses, then it is cheaper to have the parts on stock. Also, it makes sense to get external companies in for the larger tasks and to do the smaller ones internally; this also keeps know-how within the company. It is, in general, very helpful to draw a clear boundary between what is large and small but this can only be done relative to the individual plant.

A principal discovery is that the data to maintenance requests, jobs and reports is often incomplete in that entries may be missing altogether or fields in an entry are empty. Other such items are not correctly reported work reports. This is often not discovered in most enterprises because most of the data is not analyzed for internal consistency. Once a self-benchmarking study is undertaken, the data is put into relation to itself and all inconsistencies become apparent. This helps an enterprise by directly improving its data management.

Throughout the data, there will also be anomalies. These are very unusual for some statistical reason. For example, they took far too long, cost far too much effort or material and so on. This outlier detection that is intrinsic to data-mining is very helpful to find the items that were either truly unusual (and should therefore be treated specially) or that were incorrectly input into the database.

Unifying different tables in database frequently also brings up inconsistencies. This could be that the measures as such lead to a different total cost as the summary of material and personnel. This should not be but if the data is kept in different unconsolidated tables, then this may arise. It is then vital which numbers will be taken as the final ones. Such effects raise the question whether the database is correct enough for at least one of the points of view to be correct.

Self-benchmarking is a method that compares the present state of a production plant to a past state of the same plant and draws similar conclusions to a normal benchmark study. It concerns itself mainly with the costs and tries to identify potentials for cost savings in the future. We have examined this on the example of maintenance costs in several chemical plants.

We come to several main conclusions: (1) Self-benchmarking is significantly faster and cheaper than normal benchmarking, (2) self-benchmarking yields useful results that allow the plant to reduce costs in the future and find the areas of potential, (3) data management is crucial as a basis for the study and will be improved through the first such study, (4) several useful insights into the operation of the business are possible that will allow relevant changes to be made for the better, (5) this type of study can be automated and be displayed regularly so that the plant is always aware how it is doing.

4.9 Case Study: Financial Data Analysis for Contract Planning

Co-Authors: Hans Dreischmeier, Vestolit GmbH & Co. KG
Kurt Müller, Vestolit GmbH & Co. KG
Markus Ahorner, algorithmica technologies GmbH

In the maintenance department of a large chemical facility with several plants, we are faced with the problem of budget planning for the future. In this particular case, we are analyzing the last ten years. The first five were spent by doing the maintenance in house and the next five were spent with the maintenance outsourced to a service provider. This service provider has a fixed-fee contract so that a single yearly price pays for all necessary maintenance efforts required to have the plant running at a certain availability (or better). We therefore have all relevant data, both technical and economic, only for the first five years. For the second five years, we have only the technical data.

The management team of both the plant and the service provider feel now that this fixed price was too low and must be re-negotiated. There are three possible scenarios for the future: (1) in-source the maintenance again, (2) make another fixed-fee contract with some other value of the fee and (3) continue outsourcing but agree on a catalog of individual services and prices and thus remove the obligation of a minimum availability from the service provider. The question is, which alternative is better (and why)?

We note in passing that "better" is not the same as "cheaper" for the problem has more facets than money alone. The problem of employee (and thus knowledge) rotation, the dichotomy of risk and reward, the tension between saving on maintenance by quickly degrading the equipment, the security and financial planability in a fixed cost environment versus the potential saving in a variable cost environment and so on add interesting features. We must essentially consider that each of these risk factors has an associated price and consider the best solution in this complex of possibilities.

The inherent issue is that all the important events, technical and risk factors, will only happen with a certain probability. We cannot be certain of anything and so our approach is necessarily probabilistic.

We will view the problem from three angles: plants, equipment categories and employees. The whole budget as a single ticket per year is divided into each of several plants, into each of several equipment categories and each of several employee categories.

The purpose of doing this is that we can anticipate, from our technical records of failures, a certain failure frequency in each plant and each equipment category. We know, for reasons of technical skill and education, which employee category is needed for a particular failure and we have trade union tariffs for these categories.

The essential underpinning of this approach is the simple observation that the price of a single maintenance measure forms a tightly clustered normal distribution if we specify these three categories. We have been able to determine this fact from the first five years of experience in which we have all data. This means that we must merely know the *number* of future failures in order to compute a budget. We then use our technical records of all ten years to project the number of failures in each category into the future.

Please note that we must calculate a future value for the number of failures in any given tuple of (plant, equipment, employee) given only ten past values, one per year. The dearth of data restricts the mathematical sophistication with which we may treat the problem. Depending on the configuration, we will use linear or simple polynomial regression to make this projection; see section 5.3.1 for details.

Naturally the future is not the same as the past. Equipment gets older and thus fails more frequently, equipment is exchanged and thus fails less frequently, employees are trained further and become better or change categories, plants are in higher demand one year and less the next. Such factors must be taken into account in our future projections of the number of failures.

The fact that the normal distribution of price has a standard deviation provides us with a natural measure of the uncertainties involved in our approach. We may thus compute the future costs and the uncertainties in this estimate.

What do we end up with? We end up with a diced cube. The three dimensions are plant, equipment and employees. The dicing in each dimension is the categorization. Each dice in the cube now contains two numbers: The expected number of failures to be treated and the expected cost of this with their respective uncertainties. The fact that we have 20 plants, 128 equipment categories and 4 employee categories means that we have a total of 10240 boxes to fill.

In order to make this mass of numbers a little more transparent, we may plot slices through this cube as a contour plot. See figure 4.8 for one example of this. The horizontal axis enumerates the equipments and the vertical axis the plants. This is thus a slice for a particular employee category. The tetrahedra on the plot indicate the cost structure. The more pronounced the shape, the higher the cost. With some knowledge of the underlying plant, this plot can be interpreted more readily than a long list of numbers and structures can be observed. As this requires specific knowledge of the plant in question, we cannot discuss these conclusions here. We do note however that such graphical treatment is useful in most cases to get a handle on many numbers.

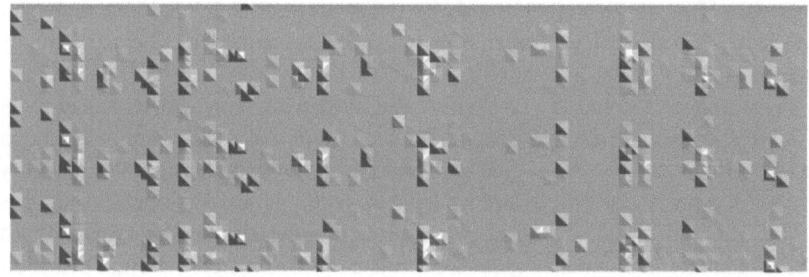

Fig. 4.8 A contour plot of the expected cost structure for one particular employee category. The cost is displayed by the tetrahedra coming out of the plane of the page, the horizontal axis is the equipment categorization and the vertical axis is the plant categorization.

This done, we first check this for plausibility in that we compute the expected cost of maintenance in the second five years. We do not have the breakdown because the activities are outsourced but we do know the total cost as this is reported back from the service provider. It so happens that we are right. We have thus confirmed a known global number but we have done so by aggregation of a great many details. This verifies that the approach is sound.

In this case, an interesting deviation was found. It was determined that the cost of maintenance should have been higher in the last year than it actually was. If we believe our model (and we should because it got the previous four years right), then we would have to assume that some maintenance measures were either not done or done less seriously. In the knowledge that the fixed-fee is lower than the actual cost, we must attribute this to an effort of saving money from maintenance. This however is paid for by an increasingly fast aging of the equipment. The bill of maintenance is thus only transferred from now into the future – at an unknown but high technical interest rate.

Thus we may project the computation into the future and we obtain a number and an uncertainty

This result is now the expected true cost of providing maintenance to the plant such that it is kept up in fair condition. We are thus not degrading the plant in order to save maintenance money in the short term, neither are we over maintaining. If we were to in-source the effort, this number would be the price tag. But we need to keep in mind that this price tag comes with an uncertainty as the future may bring a different failure profile than the average one assumed.

Should we wish to continue outsourcing we need to negotiate between the problems of underpaying (and then living with equipment maintained at the limit) and overpaying (and then loosing money). The fixed-fee must be higher than the above computed price to take into account the risk taken by the service provider in promising a service of unknown magnitude and in order to take into account the benefit obtained by being able to plan a specific cost. As we have computed an uncertainty figure alongside the price, it would be reasonable to set the fixed-fee price at the computed cost plus the full uncertainty.

If we are to outsource in the form of an itemized list, then we can also provide this list: It is the very cost projection per category of plant, equipment and employee that we have used above. Plus a certain margin, of course, for factors such as organization, management and liability.

After thorough analysis of the pros and cons of this approach, we emerge with a solid projection of actual cost in various dimensions and categories that provides a good basis for planning in so far as planning is possible given that the future is always uncertain. The risk factors involved are significant however.

With any outsourcing partner, there is a cost of communication and a risk of this partner doing what may not be in the interest of the plant owner. Some of these actions may be difficult to trace and may only show effects much later in time. These costs are significant but difficult to estimate in value. Also the benefit of being able to plan a definitive budget is an uncertain benefit that is difficult to price. All in all, the outsourcing partner makes a change to the price tag that is hard to compute and thus it makes a change mainly to the uncertainty parameter. Thus, outsourcing is itself a risk; at least in this constellation of factors.

We conclude in recommending insourcing in this scheme. It has the lowest estimated cost and the lowest estimated risks. The owner is in control and can do to the plant as the corporate philosophy thinks is wise. The lowering of the cost and risk is however "paid for" by accepting a maintenance budget dictated by failure events as opposed to a contract.

These conclusions may be drawn simply by carefully considering the data at hand, cleaning the data by putting it into sensible categories and checking the results for plausibility. The central piece of the approach was the observation that this categorization made the individual variables independent and distributed normally. We advise the reader, in similar tasks, to seek for a division of the problem into smaller chunks each of which is governed by factors that are mutually independent and intrinsically simple. In this case, we may then use this simple model to gain information about each factor and then re-assemble the factors into a bigger picture that has more complex features. However the complex features are *not* modeled explicitly but the emerge through aggregating many simple pieces. Thus we have generated information from data and then knowledge from information. Finally, we have created understanding from knowledge and solved our problem. This is the approach of divide and conquer.

4.10 Case Study: Measuring Human Influence

An insurance company sells its policies via a large network of non-exclusive agents and relies on word-of-mouth recommendations both by these agents, who also represent other companies, and by existing customers. We want to investigate how willing the agents are to recommend this particular insurance company as opposed to some other.

To this end, a representative sample among all agents was selected. To be representative, the collection of all agents must be first stratified into important categories relevant for the study, for instance region and rate of activity and so on. Only then will a random selection (within the strata) yield a representative sample. Each of these selected persons were now asked about their general satisfaction and willingness to recommend the insurer.

From similar studies, it is "known" that if agents are satisfied, then they are willing to recommend. Furthermore, this relationship is linear. Thus, satisfaction and recommendation are essentially the same variable, at least in this context. It is instructive to re-examine such statements in view of data taken in the context of a novel study. Indeed we find that this was not the case in this particular study. Satisfaction and recommendation were not as dependent as we would like to believe; the linear correlation was only between 0.6 and 0.7 depending on the grouping of the answers.

We now ask how to define the whole group of answers into a single score. First, we have asked the agents to score their own willingness to recommend between 0 and 10 where 0 is not at all and 10 is always. We will now declare the people who scored 9 or 10 as promoters, 7 and 8 as passives and from 6 downwards as detractors. Then we define the number of promoters minus the number of detractors as our net promoter score (NPS).

This single number now captures how much recommendation potential the company has in the market. It includes those who recommend actively and those who advise against the company. It is sensible to work on raising this score. There is one drawback in this however: The NPS reduces the entire study into a single number. As such this number means nothing. We must have comparable numbers in order to derive meaning from this. This comparison can come from competitors or, much more significantly, from repeating this study at regular intervals and comparing ourselves to ourselves at an earlier point.

A further drawback is that it will be difficult to discover the relationship between the NPS and any methods designed to increase it. The reason is simply that it is so difficult and time-consuming to measure the NPS just once. We cannot realistically introduce a good design of experiment and try out several scenarios in order to study the influence unless we are prepared to either wait a long time or do the various trials in the design of experiment simultaneously in different geographical regions.

We note that the definition of the NPS makes two elementary errors. First, it assumes that people can adequately express their feelings on a scale from 0 to 10. Psychological research shows that this is wrong. Human beings cannot grade themselves so finely. It is recommended to choose a scale from 1 to 5 at most; the best grading is into three groups: high, medium, low. That is approximately the expressibility of the average human being.

The second error is that we are grouping the 11 possible responses into three groups one of which includes 7 possible responses. This is not really fair to ourselves. There is considerable diversity in the detractor group and we artificially increase the number of its members by making its range of definition so large. We cannot nicely compare the people contained in the detractor and the promoter group. This is another reason to have people respond in less groups such as high, medium and low.

We find, also counterintuitively, that the satisfaction and recommendation scores do not correlate much with the business success of the relevant agents. This means that unsatisfied agents who do not recommend, are actually responsible for a significant chunk of the business. We do not have a good explanation for this effect at this time but this is not important as far as this book is concerned.

This case is to illustrate a few points in the basic design of data acquisition and the initial data screening. First, the data must be gathered in a way that makes a later analysis possible and sensible. Do not ask for too much or too many details. You may be overtaxing the person asked or may have to ask many more people in order to get significant answers. Second, define your concepts in a way that makes them comparable to other concepts. It is tempting to condense things down to a score but then this score must be interpretable in some way; generally we need comparisons for such an interpretation. Third, question any assumption or clear knowledge in your field. Even simple relationships may not hold in your case either for an intrinsic reason or, fatally, because of an error in the data acquisition process.

4.11 Case Study: Early Warning System for Importance of Production Alarms

Co-Authors: Manfred Meise, Hella Fahrzeugkomponenten GmbH
            Bernd Herzog, Hella Fahrzeugkomponenten GmbH

A production plant makes automotive parts on a production line involving many stations. Each station performs one step in the production process. At various stages along the line, we have checking equipment that perform a variety of functional tests on each part. When a part is found to be defective, it is flagged as such and no longer treated on the stations further down the line. If a part makes it to the end of the line, it is, by definition, a good part because it has passed all testing stations.

If a part is not ok, we know the first reason encountered for it be so. Clearly no process is perfect and so we must expect some scrap parts to be produced. Of course, we would like this to be a minimum and so we would like to be able to respond quickly if something changes and we are producing more not ok parts than normal. As the data comes in from production, we would like to know therefore if the likelihood of producing not ok parts due to any one reason has increased recently or not.

To determine this, we first collect all messages identifying a part as not ok over a longer history and then divide this list into groups by the not ok code, the reason for which the part was identified as scrap. Within each code group, we order the messages by time and compute the time difference between these messages. This is a list of numbers and we can compute the probability distribution (histogram normalized to an integral equal to one) over these values. We are going to compute such a histogram over two time periods. Each time period starts from the current moment and goes back in time. One time period is longer than the other. The exact length of each depends on the application.

Please see figure 4.9 for an example of these two distributions. We see the long-term distribution plotted as the solid line and the short-term distribution plotted as the dotted line. If the behavior of the not ok producing mechanism had been the same over the recent history as over the long-term history, those two distributions should be the same. However, we see a marked difference. Thus, we must conclude that the behavior has changed. In particular we can see that the probability of errors being separated by between 20 and 30 minutes has increased substantially in recent times.

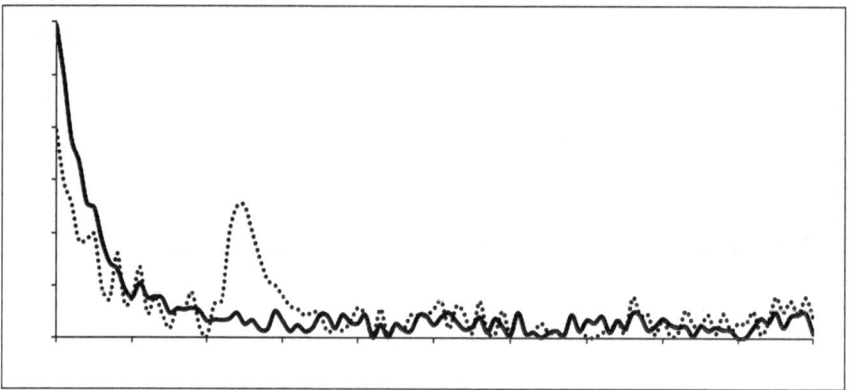

Fig. 4.9 The probability distribution measured over the long-term (solid line) compared to the same distribution over the short-term (dotted) reveals a difference. This difference is the bump between 20 and 30 on the horizontal axis. The horizontal axis measures the time difference between alarms in minutes. This image is for one particular type of alarm. In total, we have several hundred such plots for all types of alarms.

In order to automate this diagnosis, we need to be able to measure the difference between two distributions numerically. This is just what the chi-squared test is for, see section 5.2.2.3. This test gives us a measure, via the chi-squared statistic itself or via its associated significance probability, of how different these distributions are. We can then introduce a cut-off for this measure. If the distributions differ by less than this cut-off, we will consider them sufficiently similar. If they differ by more, we will signal this difference and conclude that something in the behavior of the system, with respect to this particular function test, has changed. In this case, that conclusion is provided to the operators of the production line.

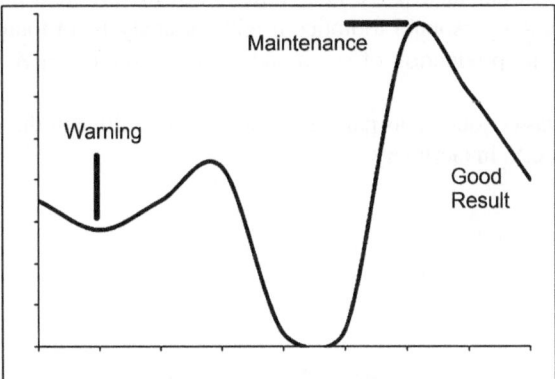

Fig. 4.10 The scrap rate due to the particular damage mechanism under study on the vertical axis is shown in dependence upon time (measured in days) on the horizontal axis. We see a rise in damaged parts up to day 7 if we ignore the production stop on days 5 and 6. Due to an early warning on day 2, a maintenance measure was implemented on day 7 that resulted in a drop of damaged parts on the following days. After this, the plant returned to normal levels, as e.g. on day 1.

In this way, we may filter the many alarms generated and provide the operators with useful feedback as to when these alarms are indeed alarming and when they are simple routine background noise.

The sensibility in doing this achieves several ends. First, it reduces the workload of the quality control team as it lets them focus on the issues that arose from changes. Second, it highlights possible problems with the production earlier in the process, as it would otherwise have taken a very pronounced change for the human team to have picked it up.

To demonstrate the effectiveness of this approach, we present one particular case in which the analysis was helpful. One particular damage mechanism is usually responsible for approximately 0.3% of scrap, e.g. day 1 in figure 4.10. That is to say if we produce 1000 parts, then 3 of these will be scrap due to this mechanism and we may or may not have further scrap parts due to other mechanisms. The plant was producing parts and the system was monitoring the scrap production of all mechanisms. On a certain day, day 2 in the figure, the system released a first warning that something was unusual with this damage mechanism. The scrap rate of this mechanism increased on the following two days. Nothing was produced on days 5 and 6 and so the scrap production was also not present. But on day 7, the scrap production due to this mechanism was already twice a strong as usual. The message from day 1 allowed a planned maintenance activity to take place on day 7. On days 8 and 9, the plant generated many fewer scrap parts and on day 10, the production settled back into its normal mode.

We observe from this example, that the statistical analysis of alarms from production extracts useful facts that are early warning signals of an impending problem and that provide actionable information. Actions can therefore be planned and implemented much earlier than if we had waited until the problem had become so severe for production personnel to notice it without analysis. In total, the analysis thereby prevents the production of scrap and increases the effective output of the plant.

This project was conducted as part of the quality committee in the society Automotive Nordwest e.V. in Germany.

Chapter 5
Data Mining: Knowledge from Data

"It is fascinating to realize that, originally, the roots of statistical analysis and data mining lie in the gaming halls of Europe. In some ways, data mining follows this heritage more closely than statistical analysis. Instead of an experimenter devising some hypothesis and testing it against evidence, data mining turns the operation around. Within the parameters of the data exploration process, data mining approaches a collection of data and asks, "What are all the hypotheses that this data supports?" There is a large conceptual difference here. Many of the hypotheses produced by data mining will not be very meaningful, and some will be almost totally disconnected from any use or value. Most, however, will be more or less useful. This means that with data mining, the inquirer has a fairly comprehensive set of ideas, connections, influences, and so on. The job then is to make sense of, and find use for, them. Statistical analysis required the inquirer first to devise the ideas, connections, and influences to test."

Dorian Pyle [107]

5.1 Concepts of Statistics and Measurement

5.1.1 Population, Sample and Estimation

Underlying all data is a reality from which that data is obtained by a process of measurement. This process will be discussed in the next section. Let us focus, for the moment, on the concept of the underlying reality.

As statistics historically found its first applications in the description of large numbers of people, the underlying reality is called the *population*. The word population may refer to a population of persons in a country or a chemical process in some particular plant. It denotes the physical reality (i.e. all theoretically possible measurements) that we wish to describe using statistical methods.

The population as such is generally too large and complex to observe as it is: There are too many persons in a country in order to ask all of them their opinion about something and there are far too many molecular interactions in a chemical

P. Bangert (ed.), *Optimization for Industrial Problems*,
DOI 10.1007/978-3-642-24974-7_5, © Springer-Verlag Berlin Heidelberg 2012

process to measure them all. Thus, we have to arrange ourselves with obtaining data about a subset of the elements in the population, a so-called *sample*.

The hope that the practically obtained description of the sample will be congruent with the theoretically existing description of the population is referred to as *estimation*. We illustrate this concept with an example. First, we believe that some physical system has a certain inherent property – let us look at the average temperature of some chemical process. Second, create a formula that calculates this property from a sample, this is called the *estimator* of the property. In this case, the formula would prescribe the summation of all temperatures measured and the subsequent division by the number of summed terms, i.e. the average. Third, we would actually apply this formula to a specific sample and arrive at the *estimate* of the property. In this case, we would arrive at a number that we would call the estimate of the average temperature.

The relationship between the actual temperature and the estimate via the average of measurements is a highly complex one. It depends, for instance, on the manner in which the sample was obtained and the size of the sample. It is beyond the scope of this book to go into this in detail. What is important to realize at this point is that a property of a system that is computed on the basis of data does not necessarily fully represent that system. The part of statistics that studies this process of obtaining a sample is called *sampling*, for more details see [53]. If we have a sample whose estimate comes close enough to the actual population property for our practical purpose, then this sample is called *representative*. It is very desirable to have a representative sample.

It is a major topic in statistics to answer the twin questions: How can we change the observation process in order to make the representation better? How good is the representation in a particular example? The first of these is hinted at in this book by giving these introductory remarks. The second is treated in cookbook fashion in section 5.2.

5.1.2 Measurement Error and Uncertainty

Whenever a measurement is taken, this measurement is uncertain and perhaps also erroneous. We will discuss these two concepts in order to make clear that a measurement is not fully representative of whatever one is trying to measure. In the context of gaining data from human beings, this seems fairly clear. Some persons change their mind, do not answer truthfully, forget things, make mistakes and so on. Thus, we must check such data for plausibility, at least, before using it for deeper analysis.

Technical data, such as a temperature measurement, also suffers from similar afflictions. Let us first discuss the uncertainty.

There are several sources of uncertainty in the context of making physical measurements. First, the sensor itself has a certain tolerance band and is not capable of measuring more accurately than some fixed amount. This is the easiest to determine as the manufacturer will label the sensor with its tolerance.

Second, the sensor will be exposed to changes in its performance based on changes in environmental conditions. An example is temperature drift where the sensor becomes more inaccurate as its own temperature rises. This already shows that uncertainty is a not a constant for a particular experiment but may, in fact, depend on the measurement result itself.

Third, the environment will influence what the sensor measures. If the sensor is dirty, clogged or covered with substance, then it will no longer measure what we desire it to measure. As we cannot know this, it becomes an uncertainty.

Fourth, the position of the sensor will reveal a number of potential sources of uncertainty as well. If the temperature sensor on a vessel in a chemical plant is outside, the temperature it measures will depend upon the weather (sunshine vs. rain and so on) much more strongly than the temperature in the vessel itself as the sensor has much less thermal inertia. Thus we would observe a variability in temperature that is unphysical. It must therefore be interpreted as an uncertainty.

These effects add up to produce a total uncertainty of a measurement. If we measure that something has a temperature of $95°C$ and this has an uncertainty of $2°C$, then we display this as $95 \pm 2°C$. This means that the actual physical process has a temperature that is in the range $[93, 97]°C$ and that it is not possible to be more precise than this.

Please note that this uncertainty has an important decision making effect: If we subsequently measure the temperature to be $96°C$, has the temperature changed? Ordinarily we would say that the temperature has increased by $1°C$ but as both measurements are uncertain by $2°C$, their ranges overlap and are thus consistent with the interpretation that the temperature of the process has not changed and that the source of this change in value is the uncertainty of the measurement. There are statistical tests to make this more precise (however only in the context of many measurements and not having only two) but this interpretational difficulty is an essential problem of measurement and will have significant practical results depending on how it is resolved!

In addition to the uncertainty, a measurement may also be in error. When a sensor breaks but still outputs a value, this value is no longer the result of a valid measurement process and thus in error. There are several sources of error.

First, any number of local effects in the process near the sensor can cause a brief spike in the measurement value of the sensor. This spike can, but does not have to be, associated with the process itself. The question whether a so-called *outlier* is a valid measurement or not is a very difficult question and can only be answered approximately by statistical methods. One popular method is to apply clustering methods and interpret those points that do not lie in large clusters to be outliers.

Second, a particular value may deviate from its historical pattern. A system is observed by a variety of sensors. As they measure properties of one process, they are all related in mutual ways. These relationships can be quantified by correlations including time-lag correlations. Should a single sensor radically deviate from this established network of relationships, we may interpret it to be in error as any physical change in the process should have affected several sensors via their relationships.

Third, the concept of systematic error is the most complex source of errors because it is not always clear that this is actually an error. A systematic error is present when that which is being measured is constantly different by the same amount from that which we want to measure. If we, for example, want to measure the average temperature of a chemical in a vat and place the temperature sensor on the wall of the vat, we are likely to measure a temperature that is permanently lower than the true average temperature. Whether this is an error (i.e. something undesirable that should be fixed) is a matter of interpretation and depends on what this value is used for. If the value is used for process stability – that is, the value itself does not matter but only the changes in the value over time – then we have nothing to worry about. However, if the value is needed for another purpose in which the value itself plays an important role, e.g. safety, then this may require an engineering change.

As soon as a measurement is taken, we need to ask whether this measurement is to be accepted at all and if we do accept it, we need to associate an uncertainty with it. Uncertainties in a measurement x are denoted by a Δ placed in front of the variable, i.e. Δx.

Suppose we have a function $y = f(x_1, x_2, \cdots, x_n)$ and the independent variables x_i have uncertainties Δx_i associated with them, then we may infer an implied uncertainty of the dependent variable y by the total differential,

$$(\Delta y)^2 = \sum_{i=1}^{n} \left(\frac{\partial f(x_1, x_2, \cdots, x_n)}{\partial x_i} \right)^2 (\Delta x_i)^2 .$$

Note carefully that various uncertainties never cancel out but always add. Also, uncertainties are weighted by the change that the independent variable causes in the dependent variable.

The fact that uncertain measurements imply an uncertainty in any subsequent computation is the major reason why these uncertainties are so important. If the measurements are made poorly, then simulation and computation may be futile efforts because the results may have such high uncertainties that they amount to no result at all.

5.1.3 Influence of the Observer

It is often the case that the very act of observation and measurement influences the system in some essential way. The placement of sensors may require certain sections of a plant to be engineered different in order to leave room for a sensor. Particularly in closed loop systems – systems in which the measurement is immediately used by an automated system to make proactive changes – the influence of measurement upon the system is clear.

This is not a problem but it is a feature that we must be aware of: The observed system is not the same as the unobserved system and so we must ideally take the nature of observation into account when designing a plant.

There are two features of this effect that requires careful thought: Non-locality and feed-back.

The fact that information gathered in one place of a process will have an effect in a far away location or at a distant time is called *non-locality*. If the control system, or the closed loop controllers, are built locally, then this may be a problem. A process in the process industry is generally spread out over macroscopic distances and may require significant timescales to come to equilibrium. It is a problem if a local controller is unaware of the effects that the local action at the current time will have in a different location or at a later time. This must be kept in mind when designing a system.

Many processes do not linearly feed material through from raw material delivery to end product dispatch but rather have some cycles of treatment, such as a chemical that may get several passes through a reactor. Due to this, a controller will affect its own operations at a later time via the element of *feed-back*. This is similar to a microphone and a speaker. As long as they are sufficiently far apart, there is no problem. When the microphone gets too close to the speaker, the random noise elements are amplified by a feed-back loop and result in the familiar loud squeak.

When designing a plant, we must be aware that the various elements do constitute a system that interacts with itself both over time and space. It is thus sensible to construct a single plant-wide simulation model that can thus take care of these systematic effects. Any local treatment cannot take care of these effects. The ignorance of local systems about global effects is both an inefficiency and a liability. The inefficiency leads to the process outputting a suboptimal yield and the liability leads to a higher accident potential than necessary.

5.1.4 Meaning of Probability and Statistics

The topics of probability and statistics in the context of the real world are much misunderstood and give rise to significant errors resulting from insufficiently careful interpretation. This section will attempt to make clear what barriers exist.

The concept of *probability* is itself not definitely certain. There are two common definitions of it that significantly influence its interpretation. The first is the *frequency* interpretation that defines a probability as the proportion of experiments that yield one result versus all possible results. The second is the *Bayesian* interpretation that defines probability as the degree of belief in something given the evidence.

The difference is elementary because the frequency interpretation must conduct trials many times and compute a ratio. In theoretical deliberations, these experiments are only thought experiments and results may be computed. In real life, we do not know systems well enough to compute anything and we must actually carry out trials for real. In this case, we must ask to what extent boundary conditions were held constant while other elements were varied and the analysis becomes difficult. On the other hand, measuring a degree of belief in a sensible, numerical and reproducible manner is not easy.

For the purposes of this book, we will use the frequency interpretation of probability. We urge the reader to take the caveats above seriously when considering to combine experimental data with theoretical knowledge in the realm of probability. Many errors derive from incorrect interpretations at the beginning or end of a computation.

Probabilities have a distribution. This is something crucial to understand for their interpretation. A probability distribution is a function $f(x)$ of some state variable x that returns the probability p that an experiment will measure the state to be in the range x to $x + dx$ where dx is an infinitesimal known from calculus. In order to be a proper probability distribution, the integral over all x must, of course, be one.

Some things are distributed uniformly: All values of x have the same probability. This is true for the faces of a coin or die – at least we expect that this is true. Many things in nature are, however, not distributed uniformly but otherwise. The normal distribution (the classic bell curve) is found often. The heights of persons or the IQ of persons are distributed normally. It is important to know the distribution of a probability in order to be able to interpret a probability correctly. Note also that (given realistic measurement accuracies) a probability always refers to the real value lying in a *range* of values as opposed to being a specific actual value.

Think for instance of a process that makes 90% of its products of type A and the other 10% of its products of type B. We now make a method that can determine the produced type with a 75% accuracy. Is this useful? Many people might say yes. It is, after all, better than 50% of pure guessing. But this is an interpretational flaw. The variable is not uniformly distributed! Type A has 90% coverage. A method that always identifies any part as being of type A will have an accuracy of 90%. Thus a method with an accuracy of 75% is poor. Any intelligent useful method must produce an accuracy higher than 90% for us to even consider it. Thus, the distribution of a probability is highly significant with respect to interpreting the results of any computation. This is particularly relevant in cases where the distribution is not immediately known or obvious.

The subject of statistics is similarly difficult to interpret. Mostly, the difficulties center around probabilities in various contexts. The most important in the industrial context is the concept of significance. This is because it is used in the common method of six-sigma analysis that is now being used in many corporations. In addition to probability, significance faces many interpretational issues.

A statement is *significant* when it is unlikely to have happened by chance or randomly. This definition already has two sources of confusion. The term 'unlikely' refers to a probability that is given as the significance level. If, for instance, something is significant at the 99% level, this means that the probability of correctly accepting the statement is more than 99%. We may make two errors in accepting statements: A *type I error* is a false positive or the act of rejecting a true statement whereas a *type II error* is a true negative or the act of accepting a false statement. The significance of a statement has a probability distribution of its own. This depends upon the statistical test used to test for it, see section 5.2. Thus, the 99% probability stated above is subject to a distribution that we did not specify.

The second term in the definition of 'significant' that causes confusion is 'by chance or randomly.' The reason is that chance events or random events also happen subject to a well-defined probability distribution. Especially in physical events of nature, these distribution may be known from the laws of physics. Do not assume that randomly means uniformly random (all events being equally likely). In nature, there are almost always preferred events and suppressed events. The distribution may be complex.

Note furthermore that the significance does not make any statement about how important, valuable or meaningful the statement is, nor that the statement points to or comes from any physical cause-effect relationship. We indeed find that there is (in a certain dataset) a significant correlation between the presence of storks and new born babies. Even though this is so, this is a significant but spurious correlation as it is not based on a cause-effect relationship. Also, it is not valuable or meaningful as we cannot draw any useful conclusions from it. This is a simple example but in industrial practice, the meaningfulness will have to be suitably questioned before action items are devised.

5.2 Statistical Testing

5.2.1 Testing Concepts

Statistical tests are methods to answer generic questions about a dataset or the relationship between two datasets. They always take the form of yes/no questions and are answered by computing a number that characterizes the dataset under investigation – called the *test statistic* – and comparing this to another number that characterizes the certainty with which we need to know the answer to the question – called the *critical value*. We then compare these two numbers and thus answer our question.

A yes/no question can, obviously, have two answers. One answer is called the *null hypothesis* and the other answer is called the *alternative hypothesis*. Depending on the result of computing both the test statistic and the critical value, we accept either one of the two hypotheses. Statistical theory desires to be particularly careful in its wording and so it speaks of not rejecting the null hypothesis instead of accepting the alternative hypothesis. We would like to be clear and practical here and thus desire to give direct answers.

We will not be going into the theory of statistical tests but rather present a practical method of applying them.

The test statistic is computed by some specific formula that depends upon the test and is computed from the data. In section 5.2.2, we will present several such formulas. This part of the test usually presents no problems. Note that the test statistic will appear twice in what follows. Once it will appear as the numerical value calculated from data. A second time, it will appear as a concept of the population.

In fact, we want to compare the data's value to the expected distribution of values for the population.

The critical value is slightly more tricky. The concept starts with realizing that the test statistic X itself is a random variable that therefore has a particular cumulative probability distribution function $P(X < x)$ giving the probability that the test statistic X, as a population concept, is less than the numerical value x for the current dataset. Supposing that $P(X < x) = 0.95$, then we know that there is a probability of 95% that the test statistic is below the measured value x and a 5% probability that it is not.

There are two points of view on how to proceed. They are equivalent but we shall discuss each separately.

First, we may begin with the value of x as provided by the test statistic evaluation on the present dataset and thus use the distribution function to compute the probability that the test statistic is below x. This value, $P(X < x)$, is then the probability with which we can accept the null hypothesis. We may now make the interpretational decision of whether we will act on the null or alternative hypotheses based on this number.

Second, we may begin with the probability p and invert the distribution function to compute the value of the test statistic that will achieve this probability, $x' = P^{-1}(p)$. Now we may compare the computed x' from the distribution function to the computed x from the empirical dataset. We accept the null hypothesis when $x < x'$. Here, we have done the interpretation once and for all in advance and specified it in the form of deciding on some value for p.

Both methods are essentially the same but they involve different steps. In general, it is very difficult to invert a distribution function, so that the second method is normally not employed except when one has pre-computed tables of the results. These tables were widely used in the pre-computer days of statistics. In modern times, the computer has taken over the analysis and now it is significantly easier to apply the first method.

The inverse probability $1 - p$ of the probability p in the second method is called the *significance level*. It denotes the largest probability with which we are prepared to mistakenly reject the null hypothesis. In other words, p is the lowest probability with which we are prepared to mistakenly reject the alternative hypothesis. Generally, p is set to 95%, 99% or 99.9% where the last is termed *virtually certain*. If p is set to such a high value, we can be sure that the answer of the test gives us an answer which we can use with confidence (in the community of most scientists). The value of p such that you feel confident is, of course, dependent upon the application and your personal disposition.

In the first method, we would take the computed probability and compare it to the significance level. If $P(X < x) > p$ we accept the null hypothesis and otherwise accept the alternative hypothesis. In the following treatment, we will be assuming this method. By using any standard statistical software, you will be able to follow this method.

In summary, this is the method:

1. Compute the test statistic and call the value x,

2. Compute the probability $P(X < x)$,
3. Choose a significance level $1 - p$, and
4. If $P(X < x) > p$, accept the null hypothesis and otherwise accept the alternative hypothesis.

In the next section, we will state a few specific tests. For each, we will give the formulas for computing the test statistic, the distribution function and the identity of the null and alternative hypotheses. This makes the above method definite apart from choosing p, which must be done in dependence upon the practical problem at hand.

5.2.2 Specific Tests

Please note that statistical theory has constructed a great many tests for various purposes. There are sometimes even several tests for a particular purpose. This book does not mean to give an exhaustive treatment. We will give a test for those questions that we consider relevant for basic data analysis in the process industry.

5.2.2.1 Do two datasets have the same mean?

For the two datasets A and B, that are thought to have the same variance, we compute the t-statistic,

$$t = \frac{\overline{x}_A - \overline{x}_B}{\left[\frac{\sum_{i \in A}(x_i - \overline{x}_A)^2 + \sum_{i \in B}(x_i - \overline{x}_B)^2}{N_A + N_B - 2} \left(\frac{1}{N_A} + \frac{1}{N_B} \right) \right]^{-1}}$$

and the distribution function

$$P(X < t) = \frac{1}{v^{\frac{1}{2}} B \left(\frac{1}{2}, \frac{v}{2} \right)} \int_{-t}^{t} \left(1 + \frac{x^2}{v} \right)^{-\frac{v+1}{2}} dx$$

where the number of degrees of freedom $v = N_A + N_B - 2$, $B(\cdots)$ is the beta function and N_A and N_B are the number of observations in either dataset.

If the two datasets do not have the same variance, the t-statistic is

$$t = \frac{\overline{x}_A - \overline{x}_B}{\sqrt{\sigma_A^2 / N_A + \sigma_B^2 / N_D}}$$

with σ_A^2 the variance of dataset A and the number of degrees of freedom are

$$v = \frac{\left(\frac{\sigma_A^2}{N_A} + \frac{\sigma_B^2}{N_B} \right)^2}{\frac{(\sigma_A^2 / N_A)^2}{N_A - 1} + \frac{(\sigma_B^2 / N_B)^2}{N_B - 1}}$$

while the distribution function remains unchanged.

The null hypothesis is that the means are the same and the alternative hypothesis is that the means are different.

5.2.2.2 Do two datasets have the same variance?

For the two datasets A and B, that are thought to have the same mean, we compute the F-statistic,

$$F = \frac{\sigma_A^2}{\sigma_B^2}$$

where $\sigma_A^2 > \sigma_B^2$. The distribution function is

$$P(X < F) = 2 - I_{\frac{v_A}{v_A + v_B F}}\left(\frac{v_A}{2}, \frac{v_B}{2}\right) - I_{\frac{v_B}{v_B + v_A F}}\left(\frac{v_B}{2}, \frac{v_A}{2}\right)$$

with $I(\cdots)$ the incomplete beta function.

The null hypothesis is that they have equal variances and the alternative hypothesis is that the variances are different.

5.2.2.3 Are two datasets differently distributed?

There are different approaches depending on the nature of the two distributions. We have to answer whether we are comparing an empirical distribution to a theoretically expected distribution or to another empirical distribution. We also have to answer whether the empirical data is in the form of binned data or available as a continuously valued distribution. Note that while binning involves a loss of information and arbitrary choice of bins, it is necessary if the dataset is in itself not a distribution and will thus convert the dataset into a probability distribution. If possible, one should not bin datasets. From these two questions, we arrive at four possibilities.

In all cases the null hypothesis is that the two sets are equally distributed and the alternative hypothesis is that they are differently distributed. Note that this test makes no statements as to how they are distributed but merely as to same or different.

One binned empirical distribution against a theory: The empirical distribution has n_i observations in bin i where we expect to find m_i observations and so we create the chi-squared statistic

$$\chi^2 = \sum_i \frac{(n_i - m_i)^2}{m_i}$$

with distribution

$$P(X < \chi^2) = \frac{1}{\Gamma(a)} \int_0^{\frac{\chi^2}{2}} e^{-t} t^{a-1} dt$$

where a is twice the number of degrees of freedom. The degrees of freedom are the number of bins used minus the number of constraint equations imposed on the theory, e.g. that the sum of expected bin counts over the theory equals that over the empirical data, $\sum m_i = \sum n_i$.

Two binned empirical distributions: When the first distribution has n_i observations in bin i and the second has m_i observations, the chi-squared statistic is

$$\chi^2 = \sum_i \frac{(\sqrt{M/N}n_i - \sqrt{N/M}m_i)^2}{n_i + m_i}$$

where $N = \sum n_i$ and $M = \sum m_i$. The distribution function is the same as above.

One continuous empirical distribution against a theory: The empirical distribution $n(x)$ is compared to a theory $m(x)$ by computing the very simple Kolmogorov-Smirnov statistic

$$D = \max_{-\infty < x < \infty} |n(x) - m(x)|$$

that is distributed according to

$$P(X < D) = 1 - 2\sum_{j=1}^{\infty}(-1)^{j-1}e^{-2j^2D^2}.$$

Two continuous empirical distributions: Two empirical distributions $n(x)$ and $m(x)$ are compared to each other. The formulas are identical to the Kolmogorov-Smirnov test above.

5.2.2.4 Are there outliers and, if so, where?

Outliers are all those measurements that do not belong into the genuine dataset. They can arise from many different mechanisms. In industrial reality, these are mostly due to faulty instrumentation, errors in the measurement chain from sensor to database or manual setup errors along this chain. As such, outliers are a big problem. Their presence can destroy or at least falsify conclusions. They must be identified and removed.

Some tests for outliers require us to specify the number of outliers present, which is of course not practical. We desire the test to tell us how many outliers there are and which data-points are outliers. The most successful test for outliers is the Rosner 1983 test [109, 110]. We compute the so called *extreme studentized deviate*

$$ESD_i = \frac{|x_{ext}(S_{i-1}) - \bar{x}(S_{i-1})|}{s(S_{i-1})}$$

for all $i \leq k$ where k is the maximum number of outliers in the dataset, S_{i-1} is the entire dataset with those $i-1$ points removed that successively are those separated by the largest difference from the mean of the remaining dataset[1], $x_{ext}(S_{i-1})$ is the point furthest away from the mean over S_{i-1}, $\bar{x}(S_{i-1})$ is the mean over S_{i-1} and $s(S_{i-1})$ is the standard deviation over S_{i-1}.

The computation of the critical levels λ_i for each ESD_i would go beyond the level of this book and we refer to Rosner's original papers [109, 110]. Supposing that $ESD_i \leq \lambda_i$ for all i, then we must conclude that there are no outliers. If some $ESD_i > \lambda_i$ for a particular i, then we look for the largest such i, i.e.

$$l = \max\{i | ESD_i > \lambda_i\}$$

and declare the l points $x_{ext}(S_0)$, $x_{ext}(S_1)$, ... $x_{ext}(S_{l-1})$ as outliers.

5.2.2.5 How well does this model fit the data?

Suppose we have experimental data (x_i, y_i) for many i. We then suppose that the function $f()$ represents the relationship between x_i and y_i to within our desired accuracy in the manner that $y_i \approx f(x_i)$. Note that it is unrealistic to expect that the equality be precise, i.e. $y_i = f(x_i)$. The reason is simply that experimental data is always subject to measurement errors and, generally, no model considers all effects.

Generally, a model $f()$ has parameters $a_1, a_2, \cdots, a_m$ that must be determined in some manner. The *fitting problem* consists of finding the values of these parameters such that the model fits the data optimally, i.e. that $\sum_i (f(x_i) - y_i)^2$ is a minimum over all possible sets of parameters. This approach is called the *least squares fitting* approach; as one attempts to find the least sum over squared terms. One typically finds this method illustrated in books in the context of fitting a straight line, $f(x) = mx + b$ with m and b being the parameters, but the approach is quite general. See section 5.3.1 for this.

The approach of least squares only prescribes the utility function (the above sum over squares) and *not* the method for finding the parameters. Finding these is a different issue and we must generally use a full optimization algorithm to do it.

Even though one will usually see least squares fitting in the context of straight lines, please note that straight lines are rare in real life situations. We almost always encounter non-linear situations and so $f()$ must be a non-linear function. The methods to find the parameters must then take account of this. The optimization methods discussed in chapter 7 can handle all such situations.

Clearly, we can use the sum $\sum_i (f(x_i) - y_i)^2$ as a score to rank different parameter assignments and choose the best one; that is the least squares method. However, when we have chosen the best one and have finished the least squares method, we are left with the questions: Is the $f()$ really representative of the data? Could a different

---

[1] We start with the entire dataset, compute its mean and select the point that is furthest from the mean on either side. We remove this point to get dataset S_1. We recompute the mean and select the point that is now furthest from the mean and proceed like this. If we are only interested in removing outliers from one end of the distribution, we must only search for points on the interesting end.

$f()$ have represented the data better? Is this least squares sum "good enough" for our practical purpose?

The least squares approach makes no statements to this effect as it considers a single $f()$ that is given to it by the human operator of the method. It is we who must choose the $f()$ and here lies the magic of modeling. *If you have chosen the modeling function $f()$ well, the modeling and optimizing are merely laborious steps that will lead you to your goal; if you have chosen $f()$ poorly, those steps will be a waste of time.*

Thus, it is elemental to verify that the model fits the data. For this purpose, we will use the chi-squared test.

First, we compute the chi-squared statistic

$$\chi^2 = \sum_{i=1}^{n} \frac{(y_i - f(x_i))^2}{f(x_i)}.$$

The probability distribution is the same as the one from section 5.2.2.3,

$$P(X < \chi^2) = \frac{1}{\Gamma(a)} \int_0^{\frac{\chi^2}{2}} e^{-t} t^{a-1} dt$$

where a is twice the number of degrees of freedom. The hypothesis that the model does indeed correctly represent the data collected is to be accepted if this probability is larger than your chosen significance level (usually 0.95 or 0.99), otherwise the model is a poor one at this significance level. Note that this method makes no statements about how to fix the problem if your model is poor. You must choose another one by yourself and try again!

5.3 Other Statistical Measures

5.3.1 Regression

The process of fitting a model to data described loosely above in section 5.2.2.5 is often called *regression*. The term regression with its generally unflattering connotations derives from the first academic use of the method: To describe the phenomenon that the descendants of tall ancestors tend to get shorter and approach the mean height of the population over the generations. This term is very commonly used for the general problem of fitting a model to data. Often the word is used in the context of straight lines. If it is used in a wider context, the literature generally speaks of non-linear regression, which then refers to the general fitting problem.

Here, we will briefly discuss a few special cases. We will present only the result, i.e. the formulas by which to compute the parameter values. If you are interested in how there were arrived at, there are many books that will describe this in detail, e.g. [62]. Recall that, with fitting, we are concerned with determining the values of the parameters from empirical data given a known model. The process of deciding on the model is generally a human decision with all of its subjective alchemical features leaving it impossible to treat in a mathematical book.

We will suppose that your data includes N observations in the form (x_i, y_i) where we will suppose that the y_i are dependent upon the x_i in some way that we wish to model. Both the x_i and the y_i may be vectors and need not be single values. The model takes the general form $y = f(x)$. It is our hope that $y_i \approx f(x_i)$ with a "good" accuracy. Presumably, we have decided what accuracy we require for our practical purpose.

We will also assume that the y_i are empirically measured quantities that have a known measurement error σ_i inherent in them, so that each measurement is actually $y_i \pm \sigma_i$. Should you wish to ignore this feature, please set $\sigma_i = 1$ for all i in the formulas below.

Should you choose a straight line, $f(x) = mx + b$, you will need to determine two parameters from the data: The slope m and the y-intercept b. This is how:

$$\Delta = \left(\sum_{i=i}^{N} \frac{1}{\sigma_i^2} \right) \left(\sum_{i=i}^{N} \frac{x_i^2}{\sigma_i^2} \right) - \left(\sum_{i=i}^{N} \frac{x_i}{\sigma_i^2} \right)^2 \tag{5.1}$$

$$b = \left[\left(\sum_{i=i}^{N} \frac{x_i^2}{\sigma_i^2} \right) \left(\sum_{i=i}^{N} \frac{y_i}{\sigma_i^2} \right) - \left(\sum_{i=i}^{N} \frac{x_i}{\sigma_i^2} \right) \left(\sum_{i=i}^{N} \frac{x_i y_i}{\sigma_i^2} \right) \right] / \Delta \tag{5.2}$$

$$m = \left[\left(\sum_{i=i}^{N} \frac{1}{\sigma_i^2} \right) \left(\sum_{i=i}^{N} \frac{x_i y_i}{\sigma_i^2} \right) - \left(\sum_{i=i}^{N} \frac{x_i}{\sigma_i^2} \right) \left(\sum_{i=i}^{N} \frac{y_i}{\sigma_i^2} \right) \right] / \Delta \tag{5.3}$$

Please note that many formulas that do not appear to be linear at first sight, actually are after the variable has been transformed, e.g.

$$y = e^{ax} + b \rightarrow y = a'z + b \text{ with } a' = e^a \text{ and } z = e^x \tag{5.4}$$

$$y = (ax)^c + b \rightarrow y = a'z + b \text{ with } a' = a^c \text{ and } z = x^c \tag{5.5}$$

$$y = ab^x \rightarrow y' = a' + b'x \text{ with } y' = \ln y, a' = \ln a \text{ and } b' = \ln b. \tag{5.6}$$

Suppose that this model is not sufficient and you would like to make things more interesting. *General linear least squares* focuses on models that are linear (in the parameters) but may take non-linear basis functions. An example is the polynomial,

$$y = a_1 + a_2 x + a_3 x^2 + a_4 x^3 + \cdots a_n x^{n-1}$$

but generally any function linear in the parameters is fine. The most general form is

$$y = \sum_{i=1}^{M} a_i X_i(x)$$

where the $X_i(x)$ are functions of x only and have *no* unspecified parameters.

To obtain the unknown parameter values a_i, we go through the following steps,

$$\mathbf{A}_{ij} = \frac{X_j(x_i)}{\sigma_i}, \text{ the } design\ matrix \tag{5.7}$$

$$\mathbf{b}_i = \frac{y_i}{\sigma_i}, \text{ the result vector} \tag{5.8}$$

$$\mathbf{a}_i = a_i, \text{ the solution vector} \tag{5.9}$$

$$\lambda = \mathbf{A}^T \cdot \mathbf{A} \tag{5.10}$$

$$\beta = \mathbf{A}^T \cdot \mathbf{b} \tag{5.11}$$

$$\mathbf{a} = \lambda^{-1} \cdot \beta. \tag{5.12}$$

Please note that the matrix $\mathbf{A}$ should have more rows than columns as we should have more data points than unknown parameters. All steps are easy except for the last, which requires us to compute λ^{-1}. As this matrix is generally large, an explicit inversion is not a good idea for many numerical reasons. The solution of the implicit equation $\lambda \cdot \mathbf{a} = \beta$ for $\mathbf{a}$ must therefore be done numerically. We suggest *singular value decomposition* (SVD) as the method of choice as it deals well with the round-off errors that accumulate.

Explaining the SVD method would go beyond the scope of this book. Any linear algebra book will explain this in painful detail, e.g. [122]. This paragraph will just explain the procedure once this decomposition is done. So, we assume that $\mathbf{A}$ has been SVD decomposed into $\mathbf{A} = \mathbf{U} \cdot \mathbf{W} \cdot \mathbf{V}^T$ where $\mathbf{W}$ is the diagonal matrix of singular values, $\mathbf{U}$ is column orthogonal and $\mathbf{V}$ is orthogonal. Then,

$$\mathbf{a} = \sum_{i=1}^{M} \left(\frac{\mathbf{U}_{(i)} \cdot \mathbf{b}}{\mathbf{W}_{ii}} \right) \cdot \mathbf{V}_{(i)}$$

where the subscript (i) refers to taking the i-th column of the respective matrix.

5.3.2 ANOVA

ANOVA is an abbreviation for *analysis of variance* that is a very popular method that is also very prone to misunderstandings. Note carefully that ANOVA is *not* a statistical test. That is, it does not answer any question with a yes or no answer, nor does it confirm or deny any hypotheses.

The method is used to investigate the effect of *qualitative* factors on a *quantitative* result. Should you have quantitative factors, you can always suppress this by grouping them into low-middle-high or good-bad groups and use ANOVA to deal with that. The principal assumption behind ANOVA is that the relationship between factors and result is *linear*. This is a crucial assumption as many relationships will be known to be non-linear in which case this method will do you no good. Historically, the method was devised as a part of the theory of the *design of experiments* in which we plan experiments before carrying them out in order to reduce the work to a minimum while gathering all required data for a certain desired future analysis.

The basic idea of ANOVA is to see if the variance of the result can be explained on the basis of the differing categories of the factors. For this, we make experiments and compute variances within and between groups of categories to see if they differ significantly. This allows conclusions as to whether the division of a factor into its categories is sensible or useful with respect to the result. For instance we may use ANOVA to compare a control group to an experimental group to see if the factor that is different really causes some observable difference in some result. As such, ANOVA is a central and commonly used tool in many experimental studies (particularly involving people). Because it is based on categorical variables rather than numerical ones, it is naturally suited towards the social studies even though it is also used in the sciences.

One use to which we may put ANOVA is the attribution of causes to effects. We may for instance find that there is a link between touching dirt and washing hands and we may find that this link has a preferred time-wise direction: Generally touching dirt occurs before washing hands. This may lead us to believe that touching dirt is a cause for washing hands. ANOVA allows us to draw this conclusion in a well-defined procedural way. As such it is useful to deal with situations in which we lack the common sense that we all have in relation to dirt and washing.

In order to use the method, the experimental data must satisfy a few conditions. If the data does not satisfy these, then the method is unusable. This is also the reason for which *you should plan your experiments, using design-of-experiments methods, before you make them*. The conditions are

1. The observations in any one group should be homogeneously distributed. This means that if we break our group of measurements into several (arbitrary) smaller groups, these should not differ in terms of their natural variation or distribution. A common example of when this does not hold is for a time series in which observations early in time are tightly clustered around a mean and then, as time passes, get less and less clustered around the mean. This group of measurements is not homogeneously distributed but heterogeneously distributed. The technical terms for this are *homoscedasticity* versus *heteroscedasticity*. There are various tests to confirm or deny this but these would go beyond the scope of this book. You will find them in many statistical books, e.g. [71].
2. The observations over the groups of any one factor are assumed to be normally distributed. If, for example, we have a factor with three groups (high-middle-low), then we would expect to find an equal number of observations in the high and low groups and a correspondingly larger amount in the middle category.
3. The observations must be independent of each other. In a time series, for example, this is definitely not true as the observation at a later time generally depends causally (often in a known way) on the observation at an earlier time.
4. If you have more than one factor, it is extremely helpful to have an equal number of observations in each combination of groups of all the factors. For example if we have two factors and each factor has three groups (high-middle-low), then we

have nine combinations of these groups (high-high, high-middle, high-low, ...)
This is called *balance*. It is not a strict requirement but life is easier if it is true.

It can easily be seen that if we are strict about these conditions (especially 2 and
3) most datasets cannot be analyzed with ANOVA. We urge you here, without any
sarcasm, to consider the principle "Never trust a statistic you didn't fake yourself"
by Winston Churchill.

Should you wish to continue beyond this point, we must now set up the presumed
model. Here we will only treat the case of a single factor. This is not very realistic
but treating more complex cases would go beyond the scope of this book, see e.g.
[87]. We then assume that the relationship between the result y and the factor is
given by

$$y_{ij} = \mu + \alpha_i + \varepsilon_{ij} \qquad \text{for } i = 1, 2, \cdots, k; j = 1, 2, \cdots, n_i$$

where y_{ij} is the result and is normally distributed within each group, μ is the mean
of the entire dataset, α_i is the effect of the i-th factor group, ε_{ij} is a catch-all for all
sorts of external random disturbances to our experiment, k is the number of groups
in our factor, n_i is the number of observations in the i-th group.

Practically, we must now compute the variance in each group σ_i^2, the means in
each group $\bar{x}_i$ and the variance of the entire dataset σ^2. Then we compute the F
statistic

$$F = \frac{n_1 n_2 (\bar{x}_1 - \bar{x}_2)^2}{(n_1 + n_2)\sigma^2} = \frac{n_1 n_2 (\bar{x}_1 - \bar{x}_2)^2}{n_1 \sigma_1^2 + n_2 \sigma_2^2}$$

to compare two groups with each other. To finish the F-test, we must look up the
value of the probability distribution of the F statistic

$$P(X < F) = 2 - I_{\frac{v_1}{v_1 + v_2 F}}\left(\frac{v_1}{2}, \frac{v_2}{2}\right) - I_{\frac{v_2}{v_2 + v_1 F}}\left(\frac{v_2}{2}, \frac{v_1}{2}\right)$$

with $I(\cdots)$ the incomplete beta function and v_1 and v_2 being the number of degrees
of freedom in each group.

The null hypothesis in this case was that the two variances are the same. If we
reject this, we must conclude that the variances are indeed different. In this case,
we would conclude that the factor does indeed have an influence upon the result.
Note that we did not say anything about the nature or strength of this influence. The
significance level at which this can be claimed is important as this can be interpreted
to be the degree to which this factor can be used to explain the variation in the result.

In this case, we used the F-test and this is quite frequent in ANOVA. Even though
we ended up using an F-test, please note that ANOVA itself is not a test and involves
a lot more than a test.

For realistic cases, we would, of course, use tests with several factors but this
goes beyond the level of this treatment, which is intended more as a caveat than a
genuine introduction.

5.3.3 Correlation and Autocorrelation

The concept of correlation is very simple. When one thing is changed, does another thing change as a consequence? If yes, these are correlated. For example, if we increase the temperature in a vessel, then the pressure will rise (supposing nothing else was changed as well). Thus, these are correlated under these circumstances. In the ideal gas situation, the variation of pressure and temperature is linear and positive: If pressure rises by a factor a, then temperature will also rise (positive correlation) by the same factor a (linear correlation). The correlation is so strong that, if volume is not modified, one variable will be enough to compute the value of the other.

We measure correlation strength numerically on a scale between -1 and 1. In the above example, the correlation would be 1. If we compare the effect of studying hard on exam results, we would expect a correlation to exist and for this correlation to be positive but it certainly will not be 1 because certain people are more prone to the subject and will thus get higher grades than others with the same amount of studying. The effect of sunlight exposure on the water level in a glass is negative: As the sun continues to shine, the water level drops due to evaporation.

Two variables x and y given by a group of measurements (x_i, y_i) have a *linear correlation coefficient* or *Pearson's r* of

$$r = \frac{\sum_i (x_i - \bar{x})(y_i - \bar{y})}{\sqrt{\sum_i (x_i - \bar{x})^2} \sqrt{\sum_i (y_i - \bar{y})^2}}$$

where the over line indicates taking an average. You should not try to determine if the correlation is significant based on r, it is not suitable for that. However, if the correlation is significant, r is a good measure of its strength.

Depending on your field of inquiry, various r values are considered "good" by the community. If you are a physicist, then you would be looking for $r = 0.99$ or so to explain an effect. Dealing with experimental data from a real life situation (as opposed to laboratory conditions), you should be happy with $r = 0.8$ or so. Many studies based on questionnaires among humans will try to interpret correlations of $r = 0.4$ and sometimes lower to have sensible meaning. Sometimes working with such low correlations is necessary when the influencing factors are too many or cannot all be measured.

In our efforts to make a mathematical model of a natural phenomenon that is known via experimental data from an industrial process, we should be happy with r between 0.8 and 0.9. If we achieve an r higher than this, we should suspect ourselves of *over-fitting* the model, i.e. introducing so many parameters that the model can memorize the data instead of extrapolating intelligently. Such over-fitting is a modeler's suicide and must be avoided.

Note that the above formula is the *linear* correlation coefficient. This assumes that the relationship between the two variables is linear. This is an important restriction as few real life relationships are linear. In order to use a non-linear correlation, we must first specify the exact form of the relationship that we believe to hold. For this reason, we cannot treat this here but must be done on a case by case basis. For a quick-and-dirty check if two variables have anything at all to do with each other, the linear coefficient is a reasonable thing to compute. Just don't base any important arguments on it.

In industrial practice, we most often deal with time-series. This is a variable that depends on or changes with time. A very interesting question about such a time-series is whether we can get a feeling for future values based on past values, i.e. knowing $x(0)$, $x(1)$, $x(2)$, ..., $x(t)$ can we make a reasonable guess at $x(t+1)$ and so on? To answer this, we will be asking how a variables correlates with *itself at an ealier time*. This is called the *autocorrelation* function $R(\tau)$,

$$R(\tau) = \int_{-\infty}^{\infty} x(t)x(t-\tau)dt.$$

The new variable τ is called the *lag* of the time-series. Usually, we normalize the autocorrelation such that $R(0) = 1$.

The autocorrelation indicates the correlation of the variable with itself at different times. The value $R(\tau)$ is a measure of the influence that the value $x(t-\tau)$ has on the value of $x(t)$. Take the example of a retail business measuring its total revenue once per month. Business is mostly stable, except in December where the Christmas business doubles revenue. We will therefore observe that $R(12)$ for this time-series will be much higher than the rest of the autocorrelation function. This indicates that there is a strong cyclic behavior with time-lag 12 (measured in months due to the data taking cadence), which agrees with our expectation. You will probably also see a stronger dimple at a time-lag of 4 and 8. These will be due to the Easter business (Easter is usually 4 months after Christmas and it usually takes 8 months from Easter to Christmas). Thus, the strength of the Christmas business may be used somewhat to predict the strength of next Easter's business and also next Christmas' business. That is the point of autocorrelation. The score $R(\tau)$ can only be interpreted relatively and offers a useful indication for further investigation.

Note that autocorrelation is time directed. That is, we measure the correlation that past events have with future events. Thus, autocorrelation is a measure of the strength of predictability: Knowing a historical fact, how reliable is an estimate of future performance on its basis (relative to knowing the future fact by simply waiting for it, which is equal to 1 by normalizing the function)? Thus, autocorrelation identifies cause-effect relationships within a single variable's time-series.

5.3.4 Clustering

Suppose that you have many observations of some process and that each observation is vector of values. We may now wish to group observations into a few qualitatively distinct categories in order to generate some form of understanding about the underlying dynamics that produce the observations. A simple example is a group of people visiting a store. Each purchase action is an observation. The observation itself is a vector of purchased goods. We now want to group the many observations made over some time interval into qualitative groups, e.g. "health conscious client" or "ready made meals client." These groups can then be described both by their purchase habits (what, how often, in which combinations ...) as well as by their economic impact (what revenue, what margins ...) in order to draw conclusions about possible changes to marketing or the store.

Each cluster should be as homogeneous as possible and the clusters should be as heterogeneous between each other as possible.

The action of grouping vectored observations into phenomenological groups is called *clustering*. The manner in which observations are clustered has some elements that are common to all methods. Beyond these common elements, there are many algorithms to accomplish a clustering and it is difficult to say *a priori* which method is best. A major reason for this, in practice, is that it is frequently not clear how to define or measure what is "best" and this emerges only empirically once the results of several methods have been compared by persons with significant domain knowledge.

First, all methods require a *metric*. A metric is a method to compute a distance measure between two observations. Some types of observation (locations and the like) have an inherent sensible distance measure but others (option A instead of option B) are difficult to tie to the concept of distance. Since we are comparing vectors of values, the issue of comparing apples with pears is also a significant problem. For instance if we measure both temperature and pressure of something in a two-dimensional vector, how are we going to measure the distance between two sample vectors? The physical units in which these quantities are measured become important. For example if we measure temperature in degrees Kelvin as opposed to degrees Celsius, then the numerical values of all temperatures will be much higher. In any normal distance measurement, the temperature will then be more significant and could thus potentially skew the results. Thus, the design of the metric is an issue that must be resolved carefully in a practical application.

Second, each clustering creates so called *centers*. These are points in the multi-dimensional space defined by the observational vectors that indicate the position of the "center" of the cluster. Each cluster has a radius around this center (as measured by the metric function) and all observations within this sphere belong to that center. We may thus list the observations per center and, via descriptive statistics, arrive at a description of each group. This description can then be interpreted and thus knowledge may be derived.

Third, some observations may not lie inside any center and thus be called *outliers*. Generally these outliers are few observations that, for one reason or another, are sufficiently atypical that they do not belong to a cluster and also sufficiently few in number not to justify forming a new cluster (or several new clusters depending on the metric function) for them.

Outliers are very interesting points as they indicate an abnormal observation. As outliers skew statistical results, it is important to look at these points in detail to determine if they are indeed genuine observations. It is possible that outliers are produced by some form of error in the observation process and would then be excluded from further treatment. If an outlier is a genuine and correct observation, then it offers insights into abnormal events. This may be important for the practical application for a variety of reasons such as capacity planning, which must orient itself on the events that are maximally taxing and so, by definition, abnormal.

Fourth, the number of centers is often a crucial point. In many practical applications, a central question is: Into how many groups is it most sensible to divide the observation? The disadvantage of most clustering methods is that the number of centers must be specified by the user before clustering is begun. In practice then, clustering is re-run with several different settings for the number of centers and each result is examined for "sensibility" whatever this may mean in the practical application at hand. Simple versions of a sensibility definition include (1) a homogeneous distribution of observations within each cluster, (2) no cluster with less/more than a specified percentage of all observations, (3) less than some specified percentage of outliers and (4) a certain specified minimum distance between centers to make sure that clusters are sufficiently distinct.

The idea of k-means requires the number of centers, k, to be specified by the user. It is an optimization problem that requires the mean-squared distance of each observation to its nearest center to be minimized by moving the k centers around in the multi-dimensional space provided by the observations.

Please note that k-means clustering is not an algorithm but a problem specification. There are several algorithms that are used to accomplish the solution of the above described problem. In fact, common optimization algorithms may profitably be used for this purpose. There is a specific algorithm, called Lloyd's algorithm, that has been invented just for this purpose: (1) Assign the observations to the centers randomly, (2) compute the location of each center as the centroid of the observations associated with it, (3) move each observation to the center that it is closest to, (4) repeat steps 2 and 3 until no re-assignments of observations to centers are made.

This algorithm is simple but it finds a local minimum. In order for us to find a global minimum, this algorithm is generally enhanced by an incorporation of simulated annealing (see chapter 7).

A sample output of a k-means clustering run is shown in figure 5.1. The data is two-dimensional in this case to make drawing an image easier. In practice the number of dimensions would generally be large. The metric used here is the Euclidean metric where the straight line between two points is the shortest distance; this will be inappropriate for many practical applications.

Having gotten this output, the question is what do we do with it? Clustering means that the observations associated with a particular cluster are in some sense similar and observations associated with different clusters are in the same sense different. The "sense" indicated here is principally measured by the metric function. The metric function is however a stepping stone and not the result because it is

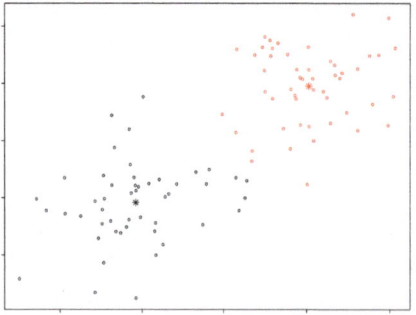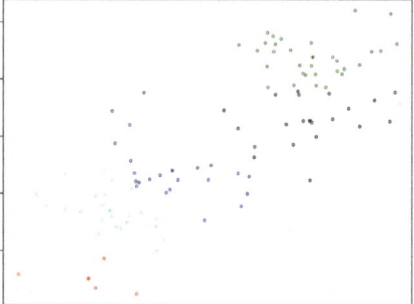

Fig. 5.1 The output of a k-means clustering with two clusters (top image) and five clusters (bottom image) on a two-dimensional dataset with the Euclidean distance metric. The centers are marked with stars and the observations with circles.

a convenient numerical measure to help the algorithm but it does not help human understanding.

What is needed at this point is to describe each cluster in such a way as to be meaningful to a human being who is charged with interpreting the data. We suggest extracting some of the following statistics, in general, for each cluster: (1) The position of the center, (2) the radius in each dimension being a measure of how large the cluster is, (3) the number of points belonging to this cluster, (4) the mean and variance over the observations in each cluster to be compared to the center position and radius as a measure of how tightly clustered the cluster really is. A comparison and critical examination of this data will allow one to discover a number of generally useful conclusions.

First, are there artifacts in the data? Artifacts are all those features of the dataset that are not intended to be there, are strange and thus to be excluded. With clustering one is likely to get artifacts from three sources: bad data (solution: better pre-processing, see chapter 4), outliers (solution: critically examine outliers and possibly exclude them) and an incorrect number of centers (solution: change k). One can determine simple artifacts by looking how clearly clusters are separated from each other. An example will illustrate the point: Two different large cities (e.g. Paris and London) are clusters of houses and are well separated as there are large tracts of virtually houseless lands in between them. However two suburbs of London are also clusters of houses but they are not well separated – their distinction is a purely administrative one and it is not immediately visible to the tourist that one section stops and another starts. Mathematically speaking thus, the division of a city into suburbs is an *artifact* that we would want to exclude (with respect to a certain viewpoint of finding clusters of houses). Clustering is there to discover meaningful qualitative differences.

Second, which dimensions principally distinguish the clusters, i.e. which attributes are sufficiently telling about an observation? This will in practice lead to the conclusion that a (hopefully small) subset of the measured parameters is sufficient to distinguish observations into their clusters. This will save effort and cost while still allowing the important conclusions to be drawn.

Third, what is the population of the clusters? If there is one cluster with many observations and other clusters with only a few each, then the situation is very different than if each cluster had approximately equally many observations. The one large cluster could be called the "normal" cluster while the others are various kinds of non-normal clusters. It depends upon the application of course, but situations with one large cluster are often correctly interpreted by saying that the large cluster acts as an attractor for the system, i.e. it is a kind of equilibrium state that a participant of the system would like to go towards. In that sense all other clusters would be pseudo-stable states that would eventually decay into the attractor. Thus, the clustering would have distinguished (pseudo-)ergodic sets from each other in the sense of statistical mechanics (see chapter 2).

Fourth, what are the application specific characteristics of each cluster? It is interesting now to compute the distinguishing features of each cluster relative to the application at hand. This depends, of course, on the nature of the problem but industrially speaking these are now parameters focusing on the major areas: safety, reliability, quality, costs, margin/profitability and the use of various resources. In this way, one can distinguish the clusters and judge them to be, in some sense, "bad" or "good" clusters. Typically this is done with respect to money using safety as a limiting criterion, i.e. we wish to maximize profitability while retaining a reasonable level of safety, reliability and quality.

Fifth, is there a dynamic in the clustering? If some of the major distinguishing features are such that they change over time, it may be possible to extrapolate a dynamic system over the clusters. This means that a participant in the system may be in one cluster at one time and in another cluster at a later time. This transition may be governed by laws that could perhaps be discovered (using other methods). In this way, a member of a "bad" cluster may be transformed into a member of a "good" cluster in some manner that, after analysis, could be understood well enough to be manipulatable.

5.3.5 Entropy

Informational entropy is a measure of the information content of a signal. High entropy means that a signal carries much information per unit of signal. If a signal is a series of letters, then the sequence "aaaaa..." is predictable and thus carries very little information. In contrast, the sequence "abcd..." carries high information content. Generally speaking, a signal source that behaves in a uniform manner will transmit a signal with near-constant entropy. While the individual measurements vary from moment to moment, the informational content of the signal is static - this is referred to as an ergodic source and is a highly desirable property for mechanical systems.

All the mechanical states that the system assumes from moment to moment and that give rise to this constant entropy are an ergodic state set. All states in one ergodic set may be interpreted as belonging to the same qualitative mode of operation. Using this, we can thus detect several qualitative states of the system over time and label these as desirable states or not.

If the system switches from one ergodic state set into another, we will observe a discontinuity in the entropy signal. This event, referred to as ergodicity breaking, is a significant event and indicates a qualitative change in operations, see section 2.5. Thus, our method of entropy tracing detects ergodicity breaking events in the history of the measurements.

Entropy can be understood by graphing a signal as a histogram. During a particular time window, the range from lowest to highest observed value is split into bins and the number of occurrences per bin over the time window is counted. Divided by the total number of measurements, this yield a probability density function that characterizes the signal over that time window, see figure 5.2 (a). The shape of this distribution characterizes the ergodic set that the system experiences during this time. If we allow some time to pass, we may detect an alternative density function, see figure 5.2 (b). The system has evolved from one qualitative state to another and this is clearly visible from the change in the density function. There has been ergodicity breaking. Entropy provides a statistical measure of this change in a single numerical quantity and allows measurement of the severity of the change.

That is only a simple example, of what the entropy method is able to detect. Most fundamental structural changes in the shape of the distribution will be detected, because they always include a change in the information.

If the entropy for a single time-window is larger/smaller than the mean of all time windows ± 2 standard deviations, then the change in entropy is big enough to let us define it as a significant change. Supposing that the distribution is normal[2] (as seen in figure 5.2 (a)), then a deviation of more than two standard deviations has a probability of less than 5%. This entropy method can be tuned to a problem by adaptively varying the number of standard deviations chosen as its detection sensitivity.

We may summarize the meaning of the abnormality score of this method: The bigger the absolute value of the abnormality score for a single time window, the higher is the difference between the distribution of values in the present time-window and the averaged distribution of all time-windows.

---

[2] If a large number of independent and identically distributed factors add up to produce a single result, then this result is approximately normally distributed - this is called the central limit theorem. Often, this theorem is used to claim that the result of almost any complex mechanism should be normally distributed. However, in practice, we find that the assumptions of the theorem (independent and identically distributed factors) hardly apply and thus the distribution is not normal. Many industrial processes have probability distributions significantly different from normal. As a result of this, we must not over interpret the claim that a data point further away from the mean than two standard deviations occurs with 5% likelihood. This is a rough guideline unless we know the identity of the distribution.

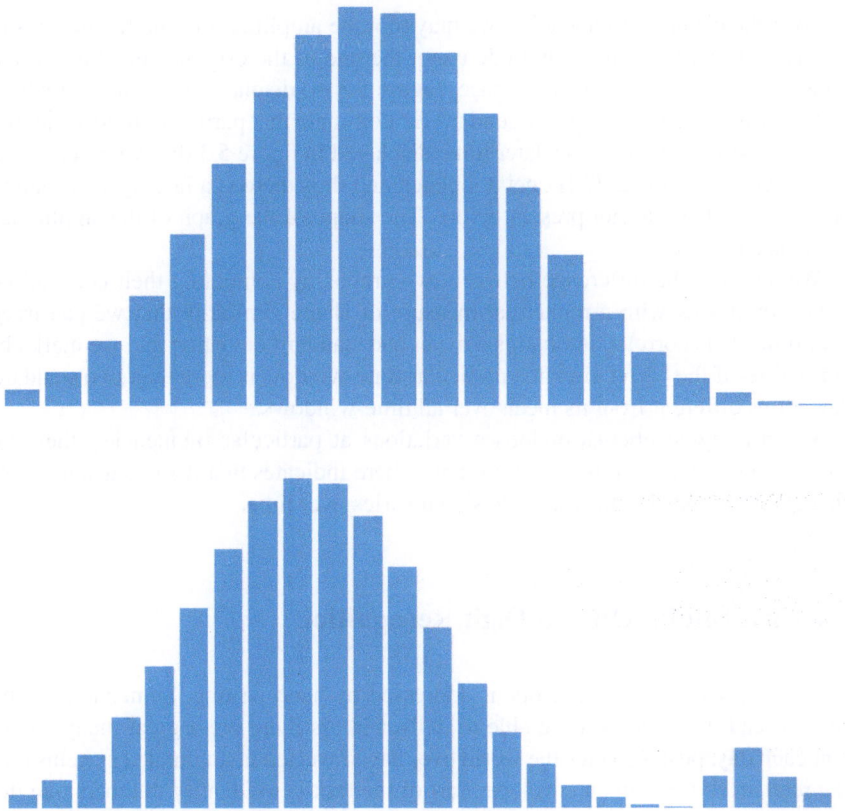

Fig. 5.2 The value distribution of a measurement before and after a qualitative change in the system that gave rise to the secondary peak on the right.

5.3.6 Fourier Transformation

The Fourier transform (FT) $\hat{f}(\zeta)$ of a function $f(t)$ is given by

$$\hat{f}(\zeta) = \int\limits_{-\infty}^{\infty} f(t)e^{-2\pi i \zeta}\,dt$$

where t is time and ζ can be interpreted to be the frequency of the signal. The transform is essentially a transformation of the basis of the coordinate system to another basis. While singular spectrum analysis (SSA) makes a linear transformation, FT makes a non-linear transformation and focuses on the frequencies with which signals change. We thus separate slow changes that occur with low frequency and fast changes that occur with high frequency.

Over the different frequencies, we may plot the amplitudes of the frequencies in the signal spectrum. This amplitude is not the one of the original signal but of the transformed signal in frequency space. Figure 5.3 (a) displays an example in which it is clear that the frequencies around 50 Hz dominate this particular time-window.

As the system evolves to a later time, displayed in figure 5.3 (b), we observe that the frequencies around 77 Hz get populated and thus there is a fast signal variation present now that was not present before. The shape of the graph of the amplitudes has thus changed.

We measure the difference between two graphs by computing their correlation. This provides us with a numerical measure of shape similarity that we can trace over time. This correlation obtains a mean and standard deviation and we mark abnormalities if the correlation at a particular time-window is more than two standard deviations different from its mean over all time-windows.

When a signal obtains or looses variations at particular frequencies, then the FT method will show this. An abnormality here indicates that there is a significant change in the speed with which the signal varies over time.

5.4 Case Study: Optical Digit Recognition

Consider a letter whose envelope is addressed by hand writing. Someone must be able to read the address to be able to deliver it. As there are a great many letters sent each day, post agencies the world over have invested in automated systems that can read an address on an envelope. Part of the task is to identify numbers like the ZIP code. Technically speaking, we are given an image of a digit and we are to say which digit the image represents. A sample of such data can be seen in figure 5.4 where we see several examples of the number six. There exists a database of about 60,000 images of digits written by about 250 different persons that we use as our dataset; this is the NIST dataset (National Institute of Standards and Technology in the USA).

In order to obtain better results, the data set is pre-processed before applying any training algorithm. One of the classical pre-processing steps is *binarization*: The image is transformed into having only black and white pixels instead of various gray levels or colors. After binarization, the data was skeletonized and thinned, see figure 5.5. However, the results obtained were worse than the results obtained using only binarization.

We will attempt to solve the problem by using a *self-organizing map* (SOM). In the terminology of section 6.4 this is a one-layer perceptron network but there is no need to skip ahead in the book as we will discuss the concept here.

We want to classify an image of a digit into the abstract categories of digits. The input of the classifier thus receives the pixels of this image and the output yields the category that this image belongs to. Thus, the SOM needs two layers. The input layer is a vector with as many elements as there are pixels in the image and so allows

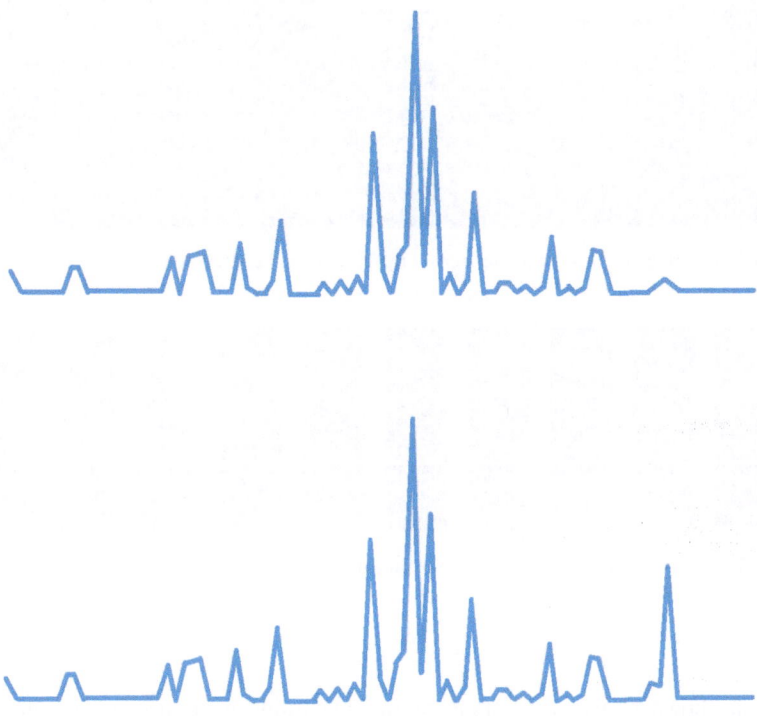

Fig. 5.3 The frequency spectrum of a signal before and after a qualitative change in the system that gave rise to the presence of frequencies around 77 Hz on the right.

for the image to be inputted into the SOM. The output layer is a vector with as many elements as there are categories; in our case of digits there are 10 such categories.

Each element of the input layer is connected with each element of the output layer and this connection has a certain strength. Thus the strengths form a matrix. In this way, every output element has a weight vector made up of the connection strength of all inputs with respect to this particular output.

When an input is presented to the network, we determine that output element whose weight vector is closest to the input vector. To measure distance, we use a metric. Most of the time, the metric is the normal Euclidean metric where the distance between (a,b) and (c,d) is

$$\sqrt{(a-c)^2 + (b-d)^2}.$$

Fig. 5.4 Several examples of the number six that have been hand written.

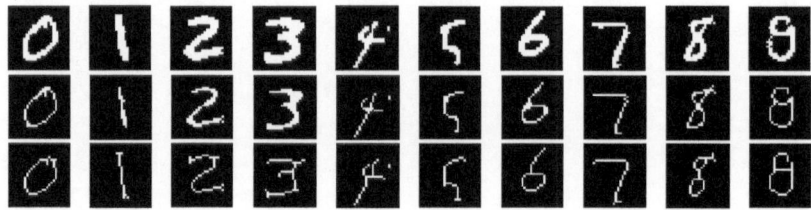

Fig. 5.5 Binarized images (on the first row) after applying thinning (on the second row) and skeletonization (on the third row)

We may however use a different metric. According to a specific algorithm, the weight vector is now adjusted slightly to reduce the distance between the input and the weight vector.

Independently of the weight vector, the output elements are considered to be locations on a map. Usually, these elements are hexagonally distributed over a two-dimensional plane, see figure 5.6 (a). The weight vectors of those output elements that are close to the specific output element just chosen are also modified by a specific algorithm. Then the next input vector is processed.

In this way, the system learns the input data and particularly learns to classify it in the form of a map. Each output element corresponds to a region in this map. We may now display this map in various graphical ways to aid the human understanding. Particularly, we may plot the distance (in the sense of the above mentioned metric) between the neighboring output elements. Close elements indicate related categories and so on. The map may of course be distorted graphically in order to reflect these distances and to give the illusion of it being a true map of something.

This way of classifying input data is particularly powerful if we have not categorized the inputs beforehand by human means. This is also an important difference. If we have input data for which we already know the output, we may teach a computer system by example as we would teach a student; this is called *supervised learning*. If we do not have this, then we must present only the input data to the computer system and hope that it will divine some sensible method to differentiate the data into categories; this is called *un-supervised learning*. For un-supervised categorization, the SOM is a very good technology. We note in passing that the accuracy of an un-supervised system is generally several percentage points worse than a supervised learning system for the simple reason that a supervised learning system has much more information (this applies for the same input data volume).

The weight vectors of each output element must be equal to some value at the start of training; this assignment is called initialization. The best results we obtained after we initialized the network with samples from the training data set. In this way the map is initially well-ordered and the convergence is faster. On Figures 5.6 (a) and 5.6 (b) is shown a map with hexagonal and rectangular topology respectively initialized from the training data.

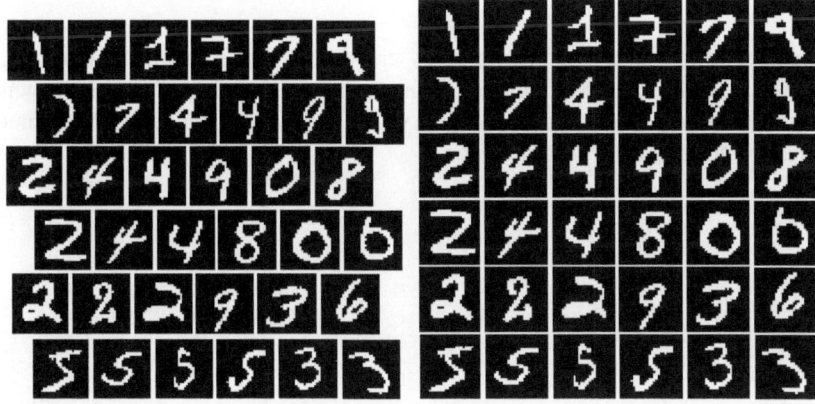

Fig. 5.6 Initialization of a 6x6 output layer with hexagonal topology on the left and rectangular topology on the right

Note that when initializing the network from the training data set, similar letters should be close to each other. For example, in Figure 5.6, the digit 1 is close to 7 since they look alike.

A crucial part of the SOM training algorithm is determining the winner node (the output element, and therefore category, assigned to the particular input image under present investigation) which is closest to the input with respect to the metric function. Thus the choice of the metric function is very important for the performance of the training. We use the so-called tangent metric and not the Euclidean metric. The reason is that it achieves substantially better results (approx. 5% difference). However, the downside of using the tangent metric is that evaluations are computationally intesive and take more time than using the Euclidean metric (about a factor 30 in time). It goes beyond the scope of this book to give the details of the tangent metric, we refer the interested reader to the literature, e.g. [72].

At first, we may observe bad performance. This may be due to a badly formulated output layer. For instance the digit 7 may be written with or without a horizontal bar and thus the number 7 is orthographically worth at least two categories. The same is true of several other digits and so we must increase the output layer from 10 to 16 as seen in figure 5.6. In this way, we may cut the error probability into half.

On a total training set of 25,000 images, we end up with an error rate of 11% on the remaining 35,000 images that were not used for training, which is not too bad given that the algorithm has no idea what we are trying to classify. The technique of re-learning is also useful: We learn and then use the final state as the starting state for another round of learning. If we make several such rounds, we can end up with a very good performance.

5.5 Case Study: Turbine Diagnosis in a Power Plant

A coal power plant essentially works by creating steam from water by heating it through a coal furnace, see figure 5.7. This steam is passed through a turbine, which turns a generator that makes the electricity. The most important piece of equipment in the power plant is the turbine.

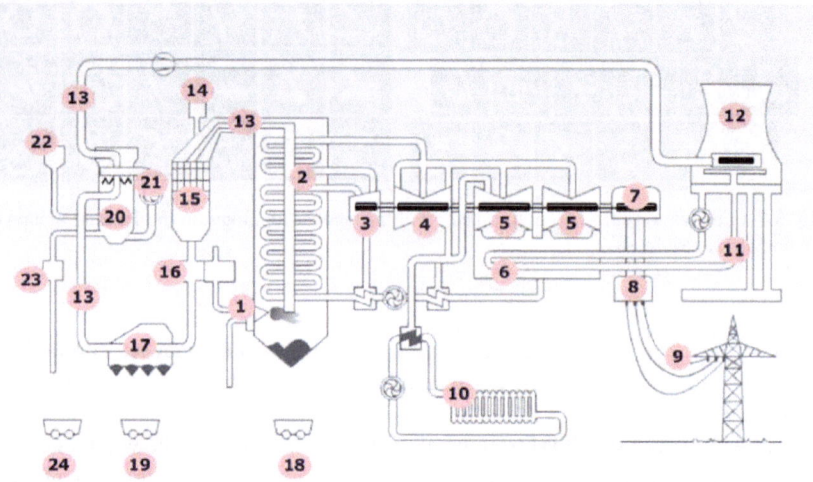

Fig. 5.7 A schematic diagram of a combined cycle power plant. The term "combined cycle" means that the power plant produces both electricity and heat for local homes. The diagram describes the combined heat and power (CHP) process in overview including the major steps: (1) Entry-point of the air, (2) boiler with water and steam, (3) high-pressure turbine, (4) mid-pressure turbine, (5) low-pressure turbine, (6) condenser, (7) generator, (8) transformer, (9) feed into power grid, (10) district heating, (11) cooling water source, (12) cooling tower, (13) flue gas, (14) Ammonia addition (15) denitrification of flue gas, (16) air pre-heater, (17) dust filter, (18) ash end-product, (19) filtered ash end-product, (20) desulfurization of flue gas, (21) wash cycle, (22) chalk addition, (23) cement/gypsum removal, (24) cement/gypsum end-product.

A turbine is a rotating engine that converts the energy of a fluid (here steam) into usable work. Figure 5.8 shows a steam turbine. Turbines are used in many other contexts such as water power plants, windmills, wind power turbines, airplane turbines and so on. We will be focusing on turbines used in standard coal-fired power plants.

The blades on the turbine have the job of actually capturing the energy and driving the central rotating shaft. They are made from steel and may be more than one meter in length. The forces at work when this machine is running are very large indeed.

Fig. 5.8 A steam turbine of Siemens during a routine inspection. Source: Siemens AG Press Photos.

Economically, a turbine is the heart piece of a power plant. All other equipment in the plant essentially caters to the turbine-generator combination because it is responsible for the conversion of energy from steam to electricity. The rest of the power plant essentially has the job of making the steam and cleaning up after itself (for example filtering the flue gas).

Thus, it is important to carefully watch the turbine for any signs of abnormal behavior. As the turbine is so important, its operations are monitored by a variety of sensors installed in key locations. The most crucial information regarding the health of the turbine is contained in the vibration measurements. All sensor output is logged in a data historian and therefore available for study.

In our case of monitoring a fleet of turbines, there are between 111 and 179 sensors measuring the condition of each turbine, which provides us with a satisfactory amount of data for the analysis.

At all times, we want to provide some automatic diagnosis that decides whether the values of the sensors are abnormal. This is the key to the analysis. Only an abnormally functioning turbine should need manual looking at and thus we want to automatically determine abnormal behavior. For this purpose, we are going to use three methods that can do this. If any set of sensors delivers abnormal values, we want to know:

1. Which method or combination of methods detected the abnormal operation?
2. How is the development of strength of the abnormalities?
3. Which sensor or set of sensors delivers the abnormal values and for how long?
4. Which sensors send abnormal values as a result of a previous abnormality?
5. When did the first/last signs of this abnormal operation appear/disappear?

We analyze time-windows from the time-series, where each time-window contains 7 days of sensor data and the time difference between two consecutive windows is 1 hour. Thus, if the result of one of the methods is that in the time-window 1 May 00:00:00 - 8 May 00:00:00 there is no abnormality, but in the time-window 1 May 01:00:00 - 8 May 01:00:00 there is some abnormality, we conclude that the observed abnormality is induced by the 1 hour difference, i.e. that it occurred on 8 May between 00:00:00 and 01:00:00.

We always use the last hour of the analyzed time interval to present the obtained results. Each analysis method delivers one value per time-window. To detect whether there was an abnormality in the analyzed time-window for a concrete method, we compute the mean and the standard deviation based on the results delivered by that method over all the time-windows. Then we compare whether the results for the current window are larger/smaller than the mean $\pm$ two standard deviations. If it is, then excess amount is called the "abnormality score" and recognized an abnormality. This means that if a sensor had the abnormality score of 4.3 on 1 May 00:00:00, then while analyzing the data for this sensor from 24 April 00:00:00 to 1 May 00:00:00, the analysis result was either greater or less than the mean $\pm (2 + 4.3)$ times the standard deviation.

We used three techniques for the investigation of the time-windows of the four datasets. All three methods use the concept of "abnormality score" as defined above and produce one abnormality score per time window and per sensor. The methods are singular spectrum analysis (see section 4.5.2), entropy (see section 5.3.5) and Fourier transformation (see section 5.3.6).

In brief, the entropy indicates if there are abnormal values in the individual measurements, the SSA indicates if there are abnormal variances in the principal component direction and the FT indicates if there are abnormal frequencies in the signal. We thus have two indicators that concern the shape of a probability density (entropy and FT) and one that concerns the size of the density (SSA). These methods concern very different indicators of abnormality and thus may or may not simultaneously detect an event. We have seven combinations of the methods for detecting an event. Each of these possibilities indicates an abnormal situation. Which methods, or which combinations of methods, respond indicates the nature of the abnormality and may assist in the diagnosis of what the cause of the abnormality may be. We describe briefly what this may mean:

1. Entropy only: We measure abnormal values but the system changes at the same speed and with the same variation as before. This must mean that the abnormal value observed is not along the principal component direction.
2. SSA only: We measure an abnormal variance but no abnormal values or frequencies. This must mean that the principal component direction changed as otherwise, we would require a significant change in the value distribution (entropy) as well. Thus, we have a qualitative change of what measurements are important for characterizing the system.
3. FT only: We measure an abnormal frequency but no abnormal values of variances. This means that the system does what it did before but it has changed the speed at which it does it.
4. Entropy and SSA: We measure abnormal values and variances but the same frequencies. The system has changed the range of its operation but not the speed at which it varies.
5. Entropy and FT: We measure abnormal values and frequencies but the same variances. As the values have changed but not the variance, this must mean that the values that changed are not along the principal component direction. Additionally, the frequencies have changed so that the inherent speed of variation has changed.
6. SSA and FT: We measure abnormal variances and frequencies but the same values. As the variance changed without having values change, this must mean that the principal component direction changed. Additionally the speed changed.
7. Entropy, SSA and FT: We observe abnormal values, variances and frequencies. The system now visits new values at new speeds and this changes the range of variation along the principal component direction. This is the most significant indication of a change in the system and these changes should be viewed as bearing the most danger.

Whether an abnormality is dangerous or not is not included in this analysis. It is likely that when a significant change in operational settings is made by the operator that an abnormality is detected even though this does not necessarily indicate a danger. However, it seems very likely that most dangerous situations would display an abnormality of at least one of the above kinds.

A zoom-in for one dataset and the SSA method shows how a particular event starts and progresses over time, see figure 5.9. Monitored over a full year, the analysis yields the result of figure 5.10. When all three methods are combined, we may compare them as per figure 5.11.

As discussed above, the combination of methods that detect a particular event lets us interpret what kind of event is taking place and thus aids the engineer in interpreting what should be done about the event. In table 5.1, we summarize, for four turbines, how many events were detected by each combination of methods. Please note that no combination is in any sense "better" than another.

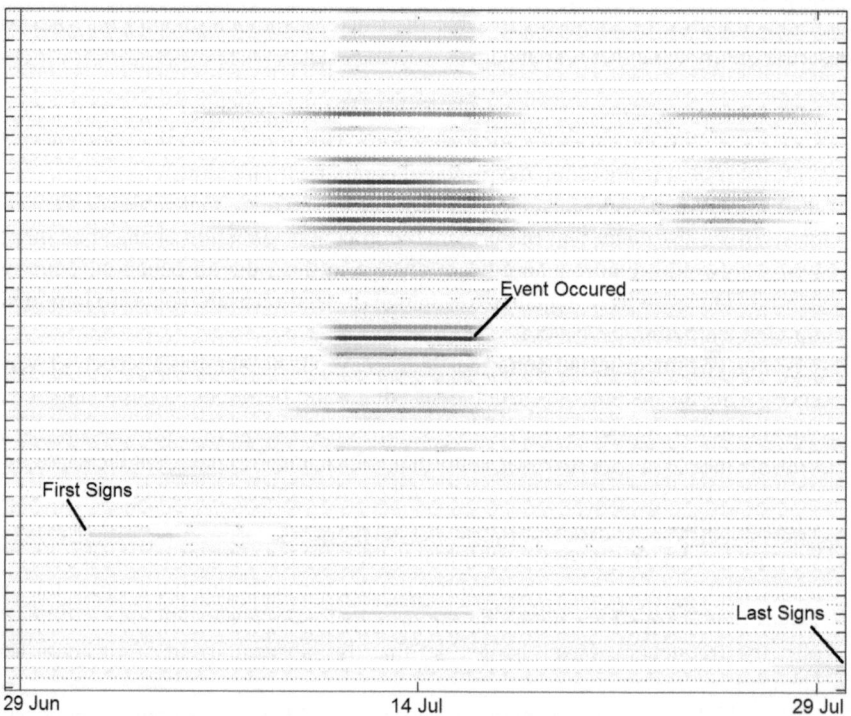

Fig. 5.9 The numbers on the vertical axis indicate the sensor that is being analyzed so that the image as a whole gives us a holistic health check for the whole turbine. We can read from the plot that the event starts with Sensor 41 on the 30th of June, then a more significant deviation is observed for sensors 122 and 151, then on the 9th of July more sensors (51) get involved and the largest abnormality (-2.09) is observed for sensor 93 on the 15th of July. For several days, abnormalities of most sensors disappear and only the sensors 122 and 151 continue deviating and start a second reaction of a smaller magnitude on the 24th of July. On the 3rd of August all abnormalities disappear.

By and large, we can see in the data that those events that have a particularly large abnormality score do tend to be detected by more than one method. The largest abnormalities are mostly detected by all three methods. On this basis, we can say that the urgency with which an event should be looked at can be proportional to the number of methods that detected it. However, this judgment is made without correlating it with the actual final outcome of the event (benign events vs. dangerous situations).

In figure 5.12, we provide a plot of the abnormality score for the events detected for one turbine.

It is also visible that the total abnormality score of an event forms an approximate exponential distribution. This is an interesting feature as this is not the outcome that would result from a large number of random interactions (central limit theorem) but rather suggests a much more structured causality. In particular, we would expect

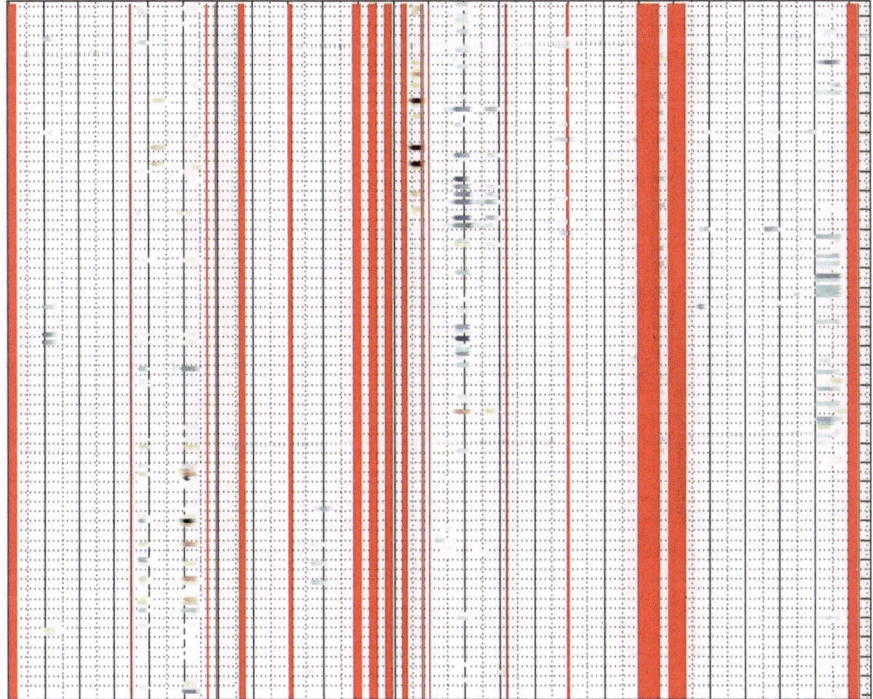

Fig. 5.10 Here we see a turbine analyzed by SSA over a full year's operation. The vertical black areas are areas where the turbine was offline. In between the offline times, we can see where the method diagnosed abnormalities. These were then analyzed by human means and appropriately responded to.

Set of Methods	Turbine 1	Turbine 2	Turbine 3	Turbine 4
SSA only	2	7	1	4
FFT only	7	4	4	1
ENT only	1	4	0	3
SSA and FFT	4	0	3	5
SSA and Entropy	3	1	1	4
FFT and Entropy	1	1	7	0
SSA and FFT and Entropy	4	6	3	3

Table 5.1 Each combination detects a particular signature of event and thus they should be seen as complementary detection schemes rather than a hierarchy. No one method dominates this table. This shows that events of all signatures do take place in the systems studied.

this outcome to result from a Poisson stochastic error source, which is present when events occur continuously and independently at a constant average rate.

Thus, we would conclude that, approximately and on average, the events detected here did not interfere with or cause each other but were independently caused. We would also conclude that whatever causation mechanism is giving rise to these

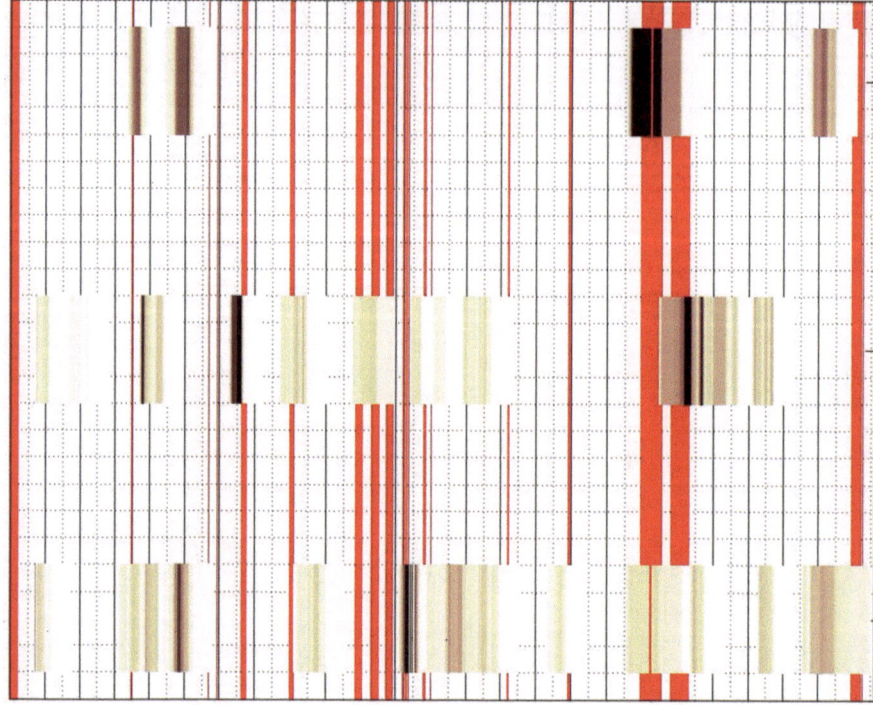

Fig. 5.11 Here we see all three methods in comparison over a whole year on one particular sensor in a particular turbine. We see broad agreement but differing opinions in the details. These differences can be interpreted as discussed in the text.

events acts at a constant rate. This means that the system does not exhibit ageing over the time period (one year) investigated here where the event rate would increase with increasing time.

We may summarize all the above findings by saying that the present methods allow the fully automatic screening of the majority of the data while finding that they indicate normal operations. For a selected small minority of the data, these methods give a clear indication at which times which sensors are abnormal, how abnormal they are and in what sense they are abnormal. This allows targeted human analysis to take place on a need basis.

5.6 Case Study: Determining the Cause of a Known Fault

Co-Author: Torsten Mager, KNG Kraftwerks- und Netzgesellschaft mbH

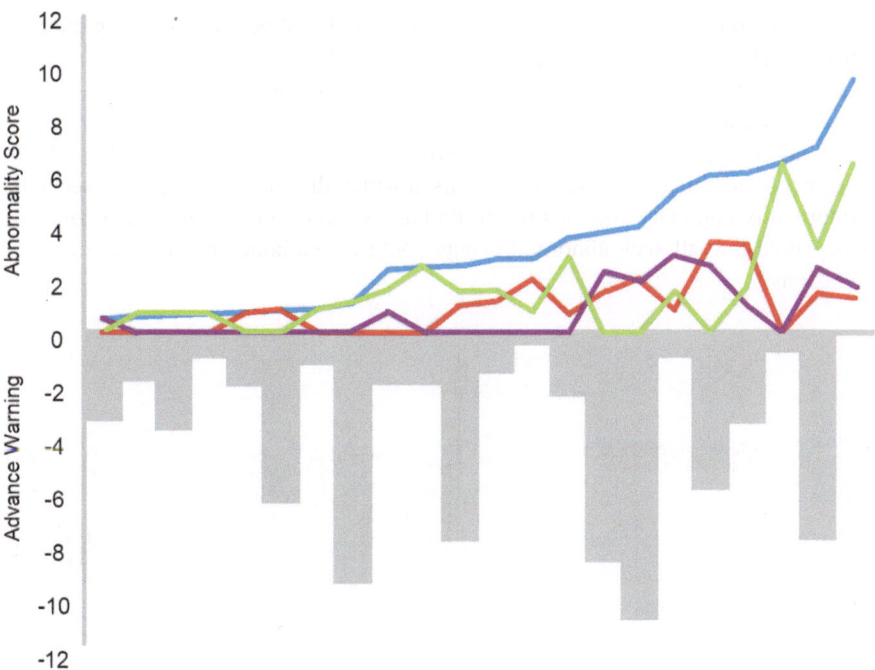

Fig. 5.12 These events are sorted by the sum of their abnormality scores of all three methods. As the score is defined similarly for each method, the numerical value of the score of one method is comparable to the score of another. We can see that the detection efficiency of the methods increases with increasing abnormality, as it should. The plot also contains the number of days before an event that an advance warning would have been possible. It is recognizable, that the first signs appeared several days in advance for most events. There are only two events where there was no sign before the event. The average advance warning time for an event was five days.

Here we take the case of turbine that has failed for a known reason. We seek the cause for the failure in both time and space, i.e. where and when did what happen to bring about the known failure? This is often important to settle issues of liability between the manufacturer, operator and insurer of the machine. It is certainly important in terms of planning what to do about it and what to do to prevent a similar case in the future. We recall that a turbine failure is significant and expensive event for the plant.

In the previous case study, we presented three methods to analyse abnormal operations for turbines. We might think that using these methods over the history of the particular turbine would reveal the point at which things went wrong.

It should be mentioned that the particular fault in question was that a single blade touched the casing and was bent. This bent blade was only detected visually upon opening the turbine months later. It was unclear to what extent the turbine would be able to continue running and so this blade was exchanged; a time-consuming and expensive process. We expect that this is an event that should be visible in the data (particularly in the vibration data) upon careful analysis.

The analysis result can be seen in figure 5.13 for one vibration measurement as an example. The three lines indicate the abnormality score of each of the three methods outlined in the previous case study. It is not important here which is which. We simply note that there are a few points at which the analysis result intersects the abnormality boundary and thus we do find abnormal events. Later manual analysis discovered that all such abnormal events could be explained by sensible operator decisions.

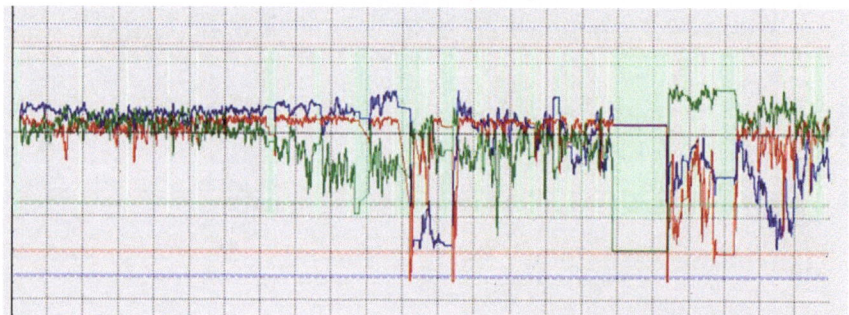

Fig. 5.13 The outcome of the analysis using the three methods of singular spectrum analysis, Fourier transform and entropy analysis are shown with respect to one vibration measurement. We observe that there are significant changes that cross boundary lines and thus indicate abnormal events.

In figure 5.14, we display the raw data in a different manner that also allows some interesting interpretations. One vibration is plotted against another in an effort to track their mutual locus. We find that they do indeed possess a well-defined mutual locus that is traversed clockwise in figure 5.14. Each point in the image represents 10 minutes of operation. A single cycle thus represents approximately 15 hours. We see on the image (on the far right) that there is a region in time lasting roughly 3 hours in which the system deviated significantly from the established locus. We can interpret this as an abnormal event.

In this particular case, however, this event was benign as it was intentionally caused by the operators. This further indicates that abnormal events do not need to be harmful. A data analysis system can detect abnormalities but would have to be given a vast knowledge to be able to interpret these as benign or harmful – at present we regard this enhancement to be impractical. We do observe however that there exist some tools that can interpret particular problematic situations on particular devices and so some work has been done in this regard [77].

In a similar fashion, we also analyzed the cases in which the turbine was not rotating at operating speed but was in various stages of cycling up or down. Also here, we were not able to find any genuine abnormality.

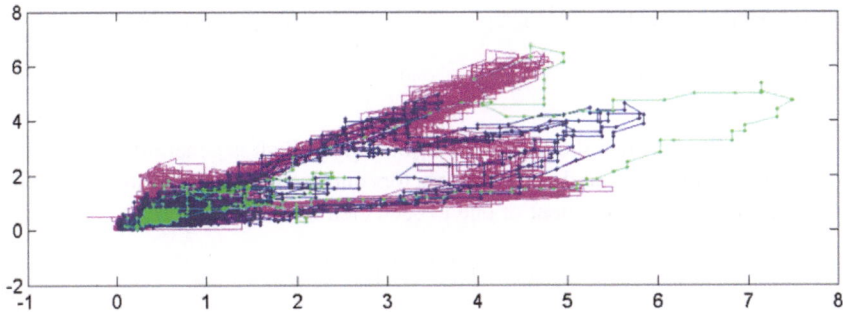

Fig. 5.14 One vibration measurement (horizontal axis) with respect to another (vertical axis). The locus is essentially a dented cycle, which (in this case) we traverse clockwise over time. We see that most of the time, the system holds a fairly well-defined locus in time. Occasionally, we do deviate from this and this represents, loosely defined, an abnormal event.

Even though this is a negative example of data analysis, it is instructive in various ways that must be taken seriously when starting such an analysis.

In particular, it is clear that there are events that cannot be seen – even upon detailed analysis – in the normal measurements of a plant. To circumvent this, we may have to install further instrumentation equipment specifically targeted at discovering such a problem. We must be aware that before we conduct a data analysis that it is possible that the feature we are looking for: (1) does not exist at all, (2) is not contained in the data we have, (3) is overshadowed by noise in the data, (4) occurs at such short timescales that it appears to be an outlier and so on. Ideally, we would have the opportunity to design a data acquisition system in order to look for a particular problem in advance. When we encounter a problem however, then we will simply have to deal with the data available and may then encounter the above challenges. The correct acquisition and cleaning of data prior to analysis is crucial for success.

We conclude this case study by observing that there are two principal explanations for the failure to find the cause:

1. The event occurred during the time period analyzed but was not visible in the data possibly because it was too short-lived.
2. The event did not occur during the time period analyzed. This would imply that the turbine was initially taken into operation in a damaged state.

5.7 Markov Chains and the Central Limit Theorem

A *Markov chain* is a sequence of numbers, each of which is a random variable, with the property that the probability distribution of any one number depends only upon the previous number in the sequence. This property is called the *Markov property.*

Thus, a Markov chain $z^{(0)}, z^{(1)}, z^{(2)} \cdots z^{(m-1)}, z^{(m)}$ has the property that

$$p\left(z^{(m+1)}|z^{(0)}, z^{(1)}, \cdots z^{(m)}\right) = p\left(z^{(m+1)}|z^{(m)}\right).$$

This probability is called the *transition probability*, which is generally a matrix and not a scalar as both $z^{(m)}$ and $z^{(m+1)}$ are vectors and we must specify the probability of transition from each element of one to each element of the other

$$T_m \equiv p\left(z^{(m+1)}|z^{(m)}\right).$$

If the transition probability T_m is the same for all m, then the Markov chain is called *homogeneous*.

The Markov property is a severe restriction and extreme simplification. It therefore allows many special properties of Markov chains to be proved. However, it also means that a Markov chain is not always suitable to model a practical situation. For that reason, we will often want to relax the Markov property and define the *p-order Markov chain* by the property that the transition probability should depend upon the prior p random variables,

$$p\left(z^{(m+1)}|z^{(0)}, z^{(1)}, \cdots z^{(m)}\right) = p\left(z^{(m+1)}|z^{(m-p-1)}, z^{(m-p-2)}, \cdots z^{(m)}\right).$$

The transition probability becomes

$$T_m \equiv p\left(z^{(m+1)}|z^{(m-p-1)}, z^{(m-p-2)}, \cdots z^{(m)}\right).$$

To model a real system by using a Markov chain, we thus need to determine the transition probabilities T_m. If we know these, and we determine the initial condition of the Markov chain (i.e. the values of the first p random variables), then we may probabilistically compute the evolution of the chain into the future and thus arrive at our model. Supposing that the initial conditions are not to be obtained from physical experiments but rather must also be calculated, we must establish the probability distribution of the initial random variables and then rely on our determination of T_m to compute the others.

As we are dealing with statistical distributions, we require a significant amount of data to be able to distinguish the various possible distributions from each other. In case of doubt, one often chooses the *Gaussian distribution* also called the *normal distribution* that looks like

$$p(z) = \frac{1}{\sqrt{2\pi\sigma^2}} e^{-\frac{(z-\bar{z})^2}{2\sigma^2}}$$

where σ is the standard deviation of the distribution and $\bar{z}$ is the mean of the distribution. The defense for this choice is often the *central limit theorem*. The users of this defense believe that the central limit theorem effectively says: If the factors leading to a particular observation are sufficiently many, the distribution of this observation will tend to the normal distribution in the limit of infinitely many factors.

Actually it states that the random variable y which is the sum of many random variables x_i (i.e. $y = x_1 + x_2 + \cdots x_m$) will tend to be distributed normally in the limit of infinitely many x's *if* the x_i are independent and identically distributed and have finite mean and variance. Please note the if clause in the previous sentence. Thus, the factors (x_i) leading to a particular observation (y) have to be independent and identically distributed, which is typically not the case as cause-effect interrelationships usually exist in the physical world. Also, the observation y is a very particular observation, namely the sum of the x's, and not some other related observation. In short, we must be very careful when using the central limit theorem to justify a normal distribution for the initial random variable in a Markov chain.

There are several popular probability distributions that optically all look the same – they have a bell shaped curve. It is also wrong to say that they are all the same and one might just as well stick with the normal distribution. In the central area of the bell, these distribution are indeed similar but they usually differ significantly on the tails (i.e. far away from the central area). Please note that in probability theory it is generally the tails that are the interesting parts because these describe the non-typical situations that will nevertheless occur. A few names of such distributions are: Cauchy, Student's t, generalized normal and logistic distribution. We will not go further into these individually.

If an incorrect distribution is chosen for $z^{(0)}$, then even a correct T_m will lead to bad results overall as everything depends on the initial condition. Thus, modeling the starting point correctly is essential for a correct Markov chain model. Moreover the correct modeling of the initial condition can only be done with sufficient data from the system under consideration. In dearth of this, we must rely on some other source such as a physical model of the situation.

5.8 Bayesian Statistical Inference and the Noisy Channel

5.8.1 Introduction to Bayesian Inference

Let us focus on a practical problem, namely optical character recognition, see for instance section 5.4 above. We may observe the system via a random variable x, the photographic image of a letter, and wish to deduce the parameter θ, the original letter, that gave rise to this observation.

We begin our excursion with the *prior distribution* $g(\theta)$ that is the probability distribution over the various possible values of the parameter. We will suppose, for the moment, that this distribution is known. We will discuss later on how it can be determined.

The observable random variable x has a conditional distribution function $f(x|\theta)$ called the *sampling distribution* that is also assumed known for all values of the parameter $\theta \in \Theta$. According to Bayes theorem, we may now compute the so-called *posterior distribution*

$$g(\theta|x) = \frac{f(x|\theta)g(\theta)}{\int_\Theta f(x|\theta)g(\theta)d\theta}.$$

Thus, we now know the distribution of the parameter given an observation. When we have made an observation, we can use this distribution to determine the probability distribution of the parameter for this particular observation.

Knowing this distribution is very useful indeed. For example, if we have an image of a handwritten letter and we determine that the probability distribution over the alphabet is such that the probability of "r" is 0.4, the probability of "v" is 0.6 and the probability of all other letters is virtually zero, then we may conclude that we have either an "r" or a "v" with high probability. We may also conclude that we are more likely to have a "v" than an "r" and we have a quantitative method to assess the degree to which it is more likely, namely 0.2 more.

We may conclude from such a distribution that our model is not yet good enough since it cannot tell these two letters apart with sufficient certainty and thus that we need to present it with more examples of these two letters to train it to be better.

Thus, it is a very useful result to have the posterior distribution. However, it depends on the prior distribution and the conditional distribution both to be known. In general these must be determined by examining the physical mechanism that gives rise to the problem. In the following subsections, we will treat the determination of these two distributions.

5.8.2 Determining the Prior Distribution

In our case, the prior distribution $g(\theta)$ is the probability that a particular letter is going to occur. We can make our life simple and deduce this from a typical piece of prose text and stipulate this as the prior distribution, see table 5.2. However, the letter frequencies are very different if we know what letter came before the current one, see table 5.3. We may, in fact, introduce a complex grammar based language model here to give an intelligent guess as to the next expected letter based on a lot of domain specific knowledge. We must make a decision what amount of knowledge must be inserted into the prior distribution for it to be good enough for our practical purpose. Noting how much complexity was added going from single letter to digram frequencies, we need to be careful before attempting to introduce a more complex model as this may create more effort than the result is worth. In the case of optical character recognition, the usefulness is however so large that several independent projects have created models of great complexity at great expense resulting in several commercial software packages.

In the case that we do not want to or cannot insert theoretical knowledge into the construction of the prior distribution, we must construct it empirically. So let us assume that we have a number of observations at our disposal $x_1, x_2, \cdots x_n$. The parameters $\theta_1, \theta_2, \cdots \theta_n$ that gave rise to these observations are unknown but we will

a	8.167%	j	0.153%	s	6.327%
b	1.492%	k	0.772%	t	9.056%
c	2.782%	l	4.025%	u	2.758%
d	4.253%	m	2.406%	v	0.978%
e	12.702%	n	6.749%	w	2.360%
f	2.228%	o	7.507%	x	0.150%
g	2.015%	p	1.929%	y	1.974%
h	6.094%	q	0.095%	z	0.074%
i	6.966%	r	5.987%		

Table 5.2 The probability of each letter in average English prose texts.

	A	B	C	D	E	F	G	H	I	J	K	L	M	N	O	P	Q	R	S	T	U	V	W	X	Y	Z
A	1	32	39	15		10	18		16		10	77	18	172	2	31	1	101	67	124	12	24	7		27	1
B	8				58			6	2			21	1	11				6	5		25				19	
C	44		12		55	1		46	15		8	16			59	1		7	1	38	16				1	
D	45	18	4	10	39	12	2	3	57		1	7	9	5	37	7	1	10	32	39	8	4	9			6
E	131	11	64	107	39	23	20	15	40	1	2	46	43	120	46	32	14	154	145	80		7	16	41	17	17
F	21	2	9	1	25	14	1	6	21		1	10	3	2	38	3		4	8	42	11	1	4			1
G	11	2	1	1	32	3	1	16	10			4	1	3	23	1		21	7	13	8				2	1
H	84	1	2	1	251	2	5		72			3	1	2	46	1		8	3	22	2		7			1
I	18	7	55	16	37	27	10				8	39	32	169	63	3		21	106	88		14	1	1		4
J					2										4						4					
K					28				8					3	3			2	1						3	3
L	34	7	8	28	72	5	1		57	1	3	55	4	1	28	2	2	2	12	19	8	2	5		47	
M	56	9	1	2	48			1	26				5	3	28	16		6	6	13	2					3
N	54	7	31	118	64	8	75		9		37	3	3	10	65	7		9	51	110	12	4	15	1	14	
O	9	18	18	16	3	94	3	3	13		5	17	44	145	23	29		113	37	53	96	13	36		4	2
P	21	1			40			7	8			29			28	26		42	3	14	7				1	2
Q																					20					
R	57	4	14	16	148	6	6	3	77	1	11	12	15	12	54	8		18	39	63	6	5	10		17	
S	75	13	21	6	84	13	6	30	42		2	6	14	19	71	24	2	6	41	121	30	2	27		4	
T	56	14	6	9	94	5	1	315	128		12	14		8	111	8		30	32	53	22	4	16		21	
U	18	5	17	11	11	1	12		2		5	28	9	33	2	17		49	42	45				1	1	1
V	15				53				19						6											
W	32	3	4		30	1		48	37			4	1	10	17	2		1	3	6	1	1				2
X	3		5		1				4						1	4				1	1					
Y	11	11	10	4	12	3	5		5		18	6	4	3	28	7		5	17	21	1	3	14			
Z					5				2					1												1

Table 5.3 The relative letter digram frequencies as measured on one English prose text. They are ordered in that the letter on the vertical axis precedes the letter on the horizontal axis in any individual diagram, so that the relative frequency of the digram "AB" is 32 and not 8.

assume that they are independent and identically distributed. Then we can group the observations into sets that are as homogeneous as possible within the set and as heterogeneous as possible between the sets. This is exactly the unsupervised clustering example that we introduced via the method of k-means in section 5.3.4 and illustrated in section 5.4 using the self-organizing map.

5.8.3 Determining the Sampling Distribution

The sampling distribution $f(x|\theta)$ is a distribution of the observable x for every value of the parameter θ. In our case, it is the distribution of all possible letter images for every possible letter. A particular letter, when it is transformed into an image can be rotated, skewed, squeezed, expanded, thinned, thickened and so on. Various mechanisms exist to make the letter image look different from a prototypical letter image – just compare your handwriting to machine typing.

Generally, we would have to have domain knowledge to construct this distribution. If this is unavailable, we may construct it empirically as we did above. We simply ask many people to write text and take this as our distribution, see section 5.4. When we cluster observations into clusters, the cluster determines the parameter and thus the prior distribution. The not quite homogeneous distribution inside any particular cluster is the sampling distribution for that particular parameter.

Please note that there is a snag: Unsupervised clustering generally produces clusters that may or may not coincide with your intended parameter. In fact, a single parameter may actually correspond to several clusters because there are various ways to distort the parameter that are very different.

In many cases, the connection between the parameter and the observation is indeed a very similar one to the case of the letter, i.e. the connection is that an idealistic concept is transformed into a physical reality via a series of mechanisms that distort the original prototype in various ways. Thus there is a channel from prototype to real object and this channel adds noise to the signal is some fashion. This noise must be removed for us to recover the original prototype.

5.8.4 Noisy Channels

There are two branches of science that deal with noisy channels. From an engineering point of view, we focus on the *noisy channel* in that we will focus on actively building a channel that has the property that we can later on remove the noise as much as possible. This view led to the construction of the television or telephone channels that are both noisy but allow the noise to be removed at the recipient's end sufficiently well to suit the user's needs. From a computer science point of view, we focus on *cybernetics* or *control theory* in that we will focus on a given noisy channel that we cannot influence and we must remove the noise as best we can. An example of this is handwriting: We cannot influence the author's handwriting and must do the best we can to recognize the intended letters via optical character recognition.

Please note carefully the difference here. In the first, we are building a channel with the noise removal problem on our mind. In the second, we are given an imperfect channel and told to remove the noise. The focus shifts from making an ideal channel to making a noise removing machine. Furthermore, when building the channel, we know to some extent the noise that the channel will add and this knowledge helps in later removal attempts but when given an existing channel, we do not generally know how it adds the noise. We will briefly present both approaches here.

5.8.4.1 Building a Noisy Channel

In constructing a channel we will want to reduce as much as possible the noise that the channel adds. For example, we want to measure the temperature in a process and convey this measurement to a process control system. The original reality is a substance that has a certain temperature. Via a sensor, cables, analog-to-digital converters, filters and so on, the information arrives at the process control system in the form of a floating-point number. This will be stored once every so often or, as is commonly done, if the value changes by more than a certain amount from the last measurement.

A noisy channel is generally characterized by a input alphabet (possible temperatures) A, an output alphabet (possible temperature readings) B and a set of conditional probability distributions $P(y|x)$. We may construct a transition probability matrix

$$Q_{ji} = P(y = b_j | x = a_i)$$

which gives the probability that the original temperature a_i is displayed as the reading b_j. A probability distribution over the input (temperatures) $\mathbf{p}_x$, which is a vector, may thus be converted into a distribution over the outputs (readings) $\mathbf{p}_y$ by multiplying it with the transition matrix like this: $\mathbf{p}_y = \mathbf{Q}\mathbf{p}_x$.

As we can influence the channel itself because we are building it, what can we do to increase our chances of correct reconstruction? That's right, we insert the same information into the channel more than once. For very important temperature measurements, the industry generally installs three sensors and the process control system records the measurement if at least two sensors agree. There are a plethora of other engineering changes that can be made to make the channel more reliable. They include procuring a good sensor with little drift and good aging properties, installing it in a location where it is not likely to get damaged, dirty or overheated, insulating the cables, setting the recording threshold low so that many value are stored and so on.

Mathematically speaking, we would control the manner in which a source signal (your voice) is encoded into electrical form over the channel (telephone) to arrive at the decoding output (speaker). We may do this via special mathematical techniques of including some, but not too much, extra information to allow the decoder to correct some of the errors that are introduced inside the channel. The adding of extra information reduces the capacity of the channel and so it is important to add the right amount. There is a large theory on exactly how much is the right amount of extra information in order to be able to decode enough of the signal for all practical purposes. Of course, this requires the "practical purpose" to be specified very accurately first. We will not go into this, as this goes beyond the scope of this book. You will find this under the headings of information theory or communication theory.

5.8.4.2 Controlling a Noisy Channel

In real life, we are more often concerned with controlling a noisy channel. That is, the channel and the encoding mechanism exist and cannot be modified. Our task is exclusively restricted to decoding, i.e. recovering the original prototype from the signal which consists of the original prototype and some noise. As the noise is generally not uniformly random but rather structured in some fashion this becomes a challenge.

Simply put, at one end we have the physical reality that we cannot directly observe, then we have a channel whose operation we do not know and at the other end we have the output that we can measure. Let us consider the channel to be a black box. Let us also suppose that we can arrange for the physical reality to be in a particular known state, at least for the purposes of experimentation if not in the final application stage. Then let us present the known physical state to the black box and observe the output. We repeat this for many trials. If we chose our inputs carefully (for example by a design of experiment, see section 3.2), then we will have observed the functioning of the channel in all important modes of operation.

What we are left with is a relationship between inputs A and outputs B. The channel is then a function $B = f(A)$. Our task is to obtain the function from the data. This is a problem of modeling that we will tackle in chapter 6.

For the moment, we will assume that the function is determined. Now, we can compute, for any input what its corresponding output will be. But we can do more. If we invert the function $A = f^{-1}(B)$, then we may compute the input for a given observed output. This finally solves the problem of determining the reality that gave rise to an observation. Of course, it is possible to construct the inverted function directly via modeling, there is no need to first model $f(\cdots)$ only to then construct $f^{-1}(\cdots)$. It is however often useful to have both orientations of the function on hand because $f(\cdots)$ is effectively a simulation of the real system and can be used for experimentation and $f^{-1}(\cdots)$ is a back trace for any output.

Suppose that we are capable of influencing the inputs A in some ways, then we can take $B = f(A)$ and ask the question: What is the value of A such that a function of the outputs $g(B)$, the so-called *goal function* or *merit function*, takes on the maximum (or minimum) value possible? This is an optimization question that we will tackle in chapter 7.

The topic of developing the model $f(\cdots)$ is effectively the development of the sampling distribution in the Bayesian inference approach. Please note that there is no contradiction between inverting the function and using Bayes' approach. The difference in method also yields a difference in output. The Bayesian inference approach does not yield a single answer (as the inversion approach does) but it yields the *distribution* of possible answers.

In practice we may be boring and desire a single answer but for more scientific work we may indeed be interested in the distribution. Also, for practical work, we would indeed be interested in the confidence (or probability) with which we may accept the single most likely answer. Perhaps the single answer of inversion is the most likely answer in Bayesian terms but there may be other answers that are sufficiently competitive that we actually cannot be confident in our conclusion.

5.9 Non-Linear Multi-Dimensional Regression

5.9.1 Linear Least Squares Regression

The word *regression* refers, in the context of statistics, to a collection of methods that infer a function describing a set of known observations. Often, regression is also referred to as *curve fitting*. A simple example is the fitting of a straight line to a set of two dimensional observations, see figure 5.15.

The simplest example is *linear regression* where we have observations of two variables (x_i, y_i) and wish to deduce a straight line $y = mx + b$ from this data. Our task, therefore, is to determine a slope m and an intercept b such that this straight line fits the observed data best. The critical word here is "best" because we will need to define very carefully what we mean by best. The classic criterion is to minimize the squared difference between model and observations, which is called the *method of least squares*. Thus, we take

$$D = \sum_i (y_i - y)^2 = \sum_i (y_i - mx_i - b)^2$$

and desire to minimize D in the space of all possible m and b.

The least squared sum is a simple function and so we can apply calculus to it, i.e. the minimum has the first derivative equal to zero,

$$\frac{\partial D}{\partial m} = \frac{\partial D}{\partial b} = 0.$$

Solving this leads to

$$m = \frac{\sum_i (x_i - \overline{x_i})(y_i - \overline{y_i})}{\sum_i (x_i - \overline{x_i})^2}, \qquad b = \overline{y_i} - m\overline{x_i}$$

where the over-line represents an average of the observed data. This method is very simple and yields a result visible in figure 5.15.

Regression lends itself to more complex questions. We may principally make matters more complex along three different directions: The observed data may be in more dimensions, the function that we hope describes the data may be non-linear and the criterion for "best" fitting may be more complex than the least squares criterion. The investigation of methods, such as the above, in the large arena opened by these three directions constitutes the field of regression.

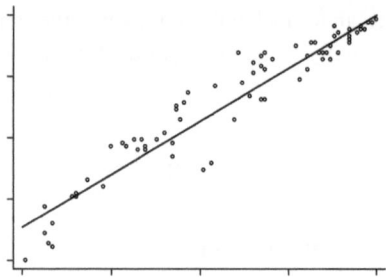

Fig. 5.15 An example of a linear regression line drawn through a set of observations in two dimensions.

5.9.2 Basis Functions

When the observations are multidimensional, we merely represent the observations by a vector $\mathbf{x}_i$. The non-linearity of the function can be complex and varied. An important possibility is to work with basis functions. An example of the basis function approach is a *Fourier series* in which we represent a function by a combination of sines and cosines,

$$f(x) = \frac{a_0}{2} + a_1\cos(x) + b_1\sin(x) + a_2\cos(2x) + b_2\sin(2x) + \qquad (5.13)$$

$$\cdots + a_n\cos(nx) + b_n\sin(nx) + \cdots. \qquad (5.14)$$

The trick in such a representation is to answer the following questions: (1) Can my function be represented in such a way in the first place?, (2) Does this series converge to the true value of my function after a practical number of terms and how many terms do I need? and (3) How do I calculate the values of the coefficients?

In the case of Fourier series, we may answer these questions: (1) If the function is square integrable, then it may be represented in this way. A square integrable function is one for which

$$\int_{-\infty}^{\infty} |f(x)|^2 dx$$

is finite. Note that this is true for nearly all functions that we are likely to meet in real life. (2) It will converge to the true value of the function at every point at which the function is differentiable. As the frequency of variation increases with each set of terms, we may terminate the series after a number of terms such that further terms would vary too rapidly for our practical purpose. In practice, we observe that this number of terms is low enough for this to be practical. Note, for instance, that music is represented (roughly) this way on a CD or in an MP3 file. (3) There are formulas for working out the value of the coefficients. We will not present them here as this would carry us too far afield.

In summary, the sines and cosines form a basis in which we may expand square integrable functions. There are other bases for other types of functions but the Fourier series is the most important one for practical applications.

Another important basis is the polynomial basis, i.e. the fact that any polynomial function can be written in terms of the powers x^n. So we may put

$$y = b_0 + b_1 x + b_2 x^2 + b_3 x^3 + \cdots + b_n x^n.$$

This is suitable if the dependency of y upon x has $n - 1$ extrema (local maxima and minima) and $n - 2$ points of inflection. To determine the b_i, we apply the same method as before. We form the least squares sum

$$D = \sum_i \left(y_i - b_0 - b_1 x_i - b_2 x_i^2 - b_3 x_i^3 - \cdots - b_n x_i^n \right)^2$$

and then put the first derivatives equal to zero

$$\frac{\partial D}{\partial b_i} = 0 \qquad \forall i : 0 \leq i \leq n.$$

This yields a system of equations that can be solved using matrix methods (e.g. Gaussian elimination) and you are left with a regression polynomial.

5.9.3 Nonlinearity

The above example of using the polynomial basis to obtain a function that is non-linear in the independent variable x is a popular method. It leaves us with the question of determining the "correct" value of n. The answer to this is complex because we first need to agree on a criterion that would allow us to objectively judge the quality of a particular choice. There are various possibilities but the least squares sum will also be useful here, i.e. select that n that minimizes D over the space of all n. This can easily be determined with a little computation.

Let us recall *Taylor's theorem* from our school days. In the simple version, it states that any $n + 1$ times differentiable function $f(x)$ can be approximated in the neighborhood of a point $x = a$ by

$$f(x) \approx f(a) + f'(a)(x-a) + \frac{f''(a)}{2}(x-a)^2 + \frac{f'''(a)}{3!}(x-a)^3 + \cdots + \frac{f^{(n)}(a)}{n!}(x-a)^n.$$

The approximation is at least as accurate as the remainder term

$$R = \int_a^x \frac{f^{(n+1)}(t)}{(n)!}(x-t)^n dt.$$

Simply put, we may approximate any reasonable function by a polynomial within some region around an interesting point.

Because this theorem applies to virtually every function that we are likely to meet in daily life, the polynomial basis is a valid approximation for virtually any function within a certain convergence interval. Of course, practically it will not be easy to determine the convergence interval but it can be determined by experimental means (deviation between polynomial and reality).

Strictly speaking, we must analyze the remainder term R and show that it approaches zero in the limit of n approaching infinity. If we can do this, the function is called *analytic* and its series expansion will converge to the function within a neighborhood around a. The neighborhood's size can be determined by determining the range of values for which the remainder tends to zero for infinite n. With closed form functions, we can compute this but with empirically determined functions, we must compare the output of the polynomial to experimental data.

To truly capture a real life case however, we need to be able to consider several independent variables. We will denote the variables by $x_{(i)}$ to distinguish them from the respective empirical measurements $x_{(i)j}$. Thus, if we want the fifth empirical measurement of the third variable we would say $x_{(3)5}$. If we now take m such variables and combine them to a maximum power of n, then the function is

$$y = \sum_{i_1=0}^{n} \sum_{i_2=0}^{n-i_1} \sum_{i_3=0}^{n-i_1-i_2} \cdots \sum_{i_m=0}^{n-i_1-i_2-\cdots-i_{m-1}} b_{i_1 i_2 \cdots i_m} x_{(i_1)}^{i_1} x_{(i_2)}^{i_2} \cdots x_{(i_m)}^{i_m}.$$

As before, we form the least squares sum

$$D = \sum_{j} \left(y_j - \sum_{i_1=0}^{n} \sum_{i_2=0}^{n-i_1} \sum_{i_3=0}^{n-i_1-i_2} \cdots \sum_{i_m=0}^{n-i_1-i_2-\cdots-i_{m-1}} b_{i_1 i_2 \cdots i_m} x_{(i_1)j}^{i_1} x_{(i_2)j}^{i_2} \cdots x_{(i_m)j}^{i_m} \right)^2.$$

To determine the parameters, we again perform the relevant partial derivatives and solve the ensuing system of coupled equations to arrive at the $b_{i_1 i_2 \cdots i_m}$.

The m will be apart from the problem specifications, it is equal to the number of independent variables that are available. However, we must be careful here: Not all available variables really do influence y (enough to matter)! Thus, we must be careful to choose only those independent variables that have a relevant and significant influence and this is a matter of some delicacy that requires domain knowledge about the source of these variables.

The n may be determined as before by performing several computations with diverse n and comparing D.

In this manner, we may deduce a non-linear model from data that will be valid at least locally. Please note that such models are inherently static as we have said nothing about time – time is not just another variable as it represents a causal correlation between subsequent measurements of the same variable and creates a totally different complication.

5.10 Case Study: Customer Segmentation

At a major European wholesale retailer, hoteliers, restaurants, caterers, canteens, small- and medium size retailers as well as service companies and businesses of all kind find everything they need to run their daily business. Every customer has a dedicated membership number and card. Due to this, it is possible to attribute every item sold to a particular customer.

Customer segmentation in general is the problem of grouping a set of customers into meaningful groups based e.g. on their profession or based on their buying behavior. In this particular case, it also allows us to trace which customers belong to these groups because we are aware of their (business) identities. This trace possibility is attempted by many other retailers via loyalty programs in which clients also allow the retailer to attach their identity to the products purchased.

Globally speaking it is interesting to find out buying patterns that can be detected in a certain group of clients. Based on a more detailed description of these groups and investigations on cause and effect for the actions of these groups, it is then possible to adjust the business model to react to such features, for example with targeted advertising such as specific products offerings to specific customers based on their purchasing habits.

Such an investigation has taken place for a dataset of all sold items in two stores over one calendar year. This included over 31 million transactions. The investigation included no particular questions to be answered and no a priori hypotheses to be confirmed or denied. The goal was to find any structures that might be economically interesting from the point of marketing to these clients.

The methods used to treat the data were diverse in nature. We used descriptive statistics, non-linear multi-dimensional regression analyses in all dimensions, k-means clustering and Markov Chain modeling. The aims were as follows:

1. Descriptive Statistics [91]: To get an overall feel for the dataset and its various sections as discovered by the other algorithms. This includes correlation analyses. In supporting the Markov chain methodology, this also includes Bayesian prior and posterior distribution analysis, which is able to tell, for example, in which order in time events happen (leads to cause and effect conclusions).
2. Nonlinear multidimensional regression [40]: To get a dependency model of the variable among each other. Expressing variables in terms of each other can lead at once to understanding and also dimensionality reduction.
3. k-means clustering [44]: To find out which purchases/clients belong into the same phenomenological group and thus determine the actual segmentation that the other methods describe.
4. Markov Chain modeling [44]: To model the time-dependent dynamics of the system and thus to find out what stable states exist.

Several descriptive conclusions are available to help with the understanding of the dataset. We present them here in a descriptive format as this is all that is required for understanding the final result. In the actual case study, these conclusions can be made numerically precise:

1. The total amount of money spent per visit is, statistically speaking, the same per visit for any particular client. Thus, in order to increase total revenues, the key is to increase customer traffic – either by getting a client to come more often or by attracting new customers.
2. Customers will generally go to the store closest to their own locations (in this study their place of business since we are dealing with a wholesaler). The probability of visiting another store decreases exponentially with distance.
3. The high seasonal business is focused in the aftermath of the summer school holidays and the preparations for Christmas. The low seasonal business is focused in the summer school holidays and during the early year post Christmas.
4. The total amount of money spent per year and per visit as well as the number of articles purchased depends highly on the type of client and the geographical region. This has a significant effect upon storage and logistics planning.
5. The majority of clients very rarely shop in the store. There is a core group of clients that shop quite regularly.
6. The products and product groups sold depend strongly on regional effects and on the visit frequency of a customer.
7. Certain products are generally bought in combination with certain other products. Thus, we may speak of a "bag of goods" that is generally bought as a whole. The contents of this bag depend upon the customer group and geography.
8. Via Bayesian analysis and Markov Chain modeling it is possible to deduce that the purchase of a certain product causally leads to the purchase of another product as an effect of the initial purchase. An example is that a purchase of fresh meat directly leads to the purchase of vegetables, cheese, and other milk products.

To summarize these conclusions, we may say that the customer behavior depends upon geography, product availability, time of the year and certain key products. It was determined that the following factors offer a significant potential in order to improve the profitability of the retail market (most significant first):

1. Individual marketing [42]: Customers tend to be interested in a narrow range of products. It is educational to cluster the customers into interest groups. We find that there are less than 10 clusters that hold a significant number of customers and that are sufficiently heterogeneous in terms of the products they offer to really divide the customers into different groups. These different interest groups could now be treated differently in some ways, e.g. by sending them advertising materials specifically targeted towards their interest group.
2. Price arbitrage: In each important product group there is a particular product that is the causal product in the group. This means that if the customer buys this product, then the customer will also buy a variety of related products in this category. This cause-effect relationship may be used to make this key product more attractive in order to boost sales in the entire product group. One way to do this is to lower the price of the key product (and raise the prices of the non-key products in the same category). It can be shown that the causal relationship is independent of price changes. However, the identity of the key product is not a universal in that there are regional differences.

3. Geography: Most sales are made to customers whose place of business is 20 to 40 minutes away from the store. At an average travel speed of 30 km/h, this is an area of approximately 940 km^2, which is comparable to the size of a moderate sized city. The wholesaler can focus his efforts, e.g. when establishing one-to-one contacts with his customers, in this area. Promotional activities in this area like billboard advertising on major roads may also be effective.
4. Time of the year: The main purchase times are March, August and pre-Christmas. The low times are January, February and summer holidays. The rest of the year corresponds to the average purchase activity. The advertising should reflect this trend, focusing on and exploiting the seasonal peaks.

Due to non-disclosure, we have presented the conclusions at such a high-level. The procedures of data-mining are able to output a quantitative presentation of these results (also with uncertainty corridors) that allows these conclusions to act as a firm basis for business decisions.

We note that these conclusions were the result of blind analysis. That is, data of 31 million transactions were given to the mining algorithms without specifying either questions or hypotheses. These algorithms output data that could be interpreted by an experienced human analyst to the above conclusions in just a few hours.

Based on these results we may now ask a number of specific questions to make the results clearer, especially when decisions have to be taken to implement changes based on these findings. We will not go into such an interactive question-answer process.

Despite the wish to know more, these conclusions are quite telling and provide valuable material for high level decision making. This illustrates very well the power of data-mining. We have converted a vast collection of data into small number of understandable actionable conclusions that can be presented to corporate management. Moreover, we have been able to do so quickly. This procedure may well be automatically reproduced monthly to track changes to customer behavior. One caveat remains however: The challenge for any data-mining approach in a "bricks and mortar" business is to translate the findings into successful operational business concepts.

Chapter 6
Modeling: Neural Networks

6.1 What is Modeling?

A *mathematical model* is a mathematical description of a system. This may take the form of a set of equations that can be solved for the values of several variables or it may take the form of an algorithmic prescription of how to compute something of interest, e.g. a decision tree to compute whether a produced part is good or bad.

It is wrong to think that all models can be written in traditional equation format, e.g. $y = mx + b$ no matter how complicated or simple the equation is. Frequently, it is far simpler to include a step-by-step recipe based on if-then rules and the like to describe how to get at the desired result. The model is thus this recipe.

Modeling is a process that has the mathematical model as its objective and end. Mostly it starts with data that has been obtained by taking measurements in the world. Industrially, we have instrumentation equipment in our plants that produce a steady stream of data, which may be used to create a model of the plant. Note that modeling itself converts data into a model – a model that fits the situation as described by the data. That's it.

Practically, just having a model is nice but does not solve the problem. In order to solve a particular practical problem, we need to use the model for something. Thus, modeling is not the end of industrial problem solving but modeling must be followed by other steps; at least some form of corporate decision making and analysis. Modeling is also not the start of solving a problem. The beginning is formed by formulating the problem itself, defining what data is needed, collecting the data and preparing the data for modeling. Frequently it is these steps prior to modeling that require most of the human time and effort to solve the problem.

Mathematically speaking, modeling is the most complicated step along the road. It is here that we must be careful with the methodology and the data as much happens that has the character of a black box.

Generally speaking modeling involves two steps: (1) Manual choice of a functional form and learning algorithm and (2) automatic execution of the learning algorithm to determine the parameters of the chosen functional form. It is important to

P. Bangert (ed.), *Optimization for Industrial Problems*,
DOI 10.1007/978-3-642-24974-7_6, © Springer-Verlag Berlin Heidelberg 2012

distinguish between these two because the first step must be done by an experienced modeler after some insight into the problem is gained and step two is a question of computing resources only (supposing, of course, that the necessary software is ready). In practice, however this two-step process is most frequently a loop. The results of the computation make it clear that the manual choices must be re-evaluated and so on. Through a few loops a learning process takes place in the mind of the human modeler as to what approach would work best. Modeling is thus a discovery process with an uncertain time plan; it is not a mechanical application of rules or methods.

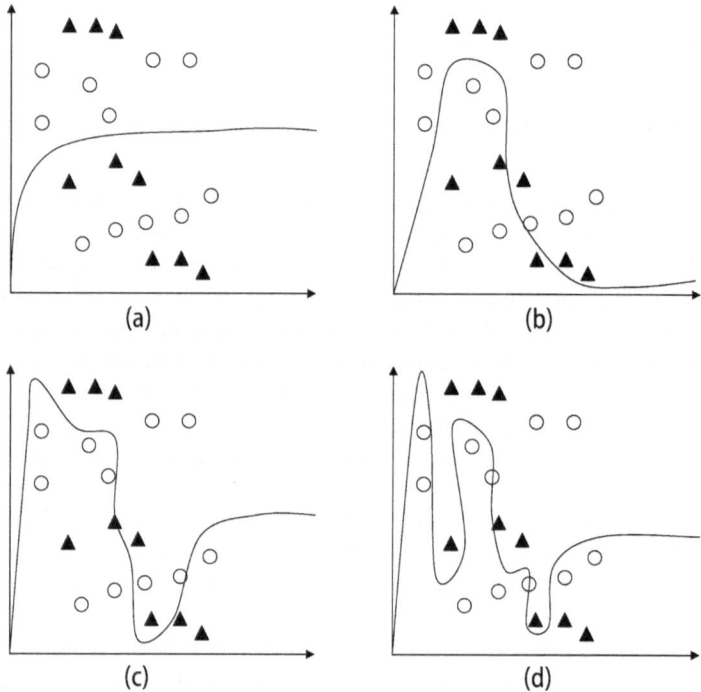

Fig. 6.1 A neural network is trained to distinguish two categories. After a few training steps (a) and two intermediate steps (b) and (c), the network settles into its final convergent state (d). In this case it may take approximately 200 iterations of a normal neural network training method to reach the final state.

Let us take the example of fitting a curve through a set of data points. First, we choose to fit a straight line, i.e. $y = mx + b$. Second, we use the linear least-squares algorithm to determine the values of m and b from the collected data. The result is the model, i.e. specific values for a and b. The algorithm used in the second step will produce the best values for the parameters that it can, given the functional form and the data. It will make an output even if a straight line is a patently poor fit for the data. Thus, it is essential to understand what type of model even has a chance to correctly model the data. Frequently model types are chosen that are provably so flexible that they can be used for nearly all data.

An example of this is the neural network that we will deal with in this chapter. It can model virtually any relationship present in a dataset. However, this statement must be taken with some salt as it is contingent upon a variety of conditions such as that the number parameters may have to be increased indefinitely for this statement to hold. That would pose a practical problem, of course, because we desire to have many fewer parameters in the model than we have data points and clearly the data points are finite in number. If we had many parameters, then the model would be no better than a simple list of data points and the modeling step would not provide us with knowledge.

It is precisely the compactification of lots of data into a functional form with a few parameters that encapsulates the knowledge gain of the modeling process. We can attempt to understand the model because it is "small." We can use the model to compute what the system will be like at points that were never measured (interpolation and perhaps extrapolation). If the number of parameters were too large, the model would merely learn the data by heart and we would loose both advantages – a situation known as *over-fitting*.

Thus, it is fair to say that *a model is a functional summary of a dataset* – it is a summary in that it encapsulates the same information in fewer numbers and it is functional in that we may use it to generate information that is not immediately apparent by evaluating it at novel points.

To make this clearer, we will look at an example in figure 6.2. This is a vibration measurement on an industrial crane. The gray jagged line displays the raw data and the dotted black line displays the cleaned data according to the methods of chapter 4. In fact, we have seen this example in figure 4.1 before. What we have added in this figure is the solid black line.

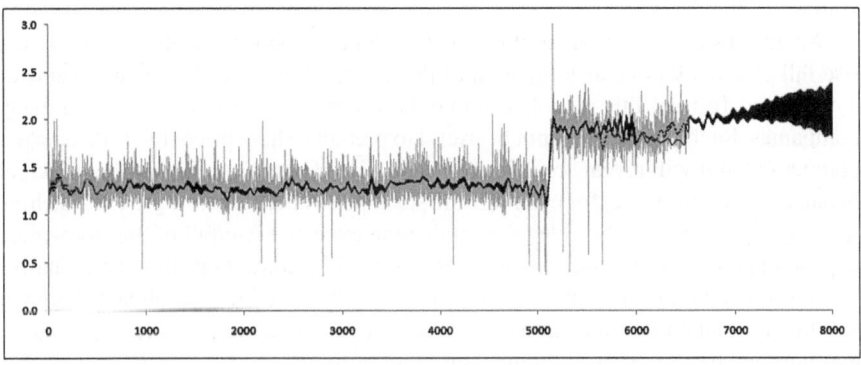

Fig. 6.2 Here we compare raw data (gray) against filtered data (black dotted) and the model (black solid) with a prediction into the future of the model.

Please note that the input data (raw as well as denoised) lasts from time zero to time 6500 minutes. This data is provided to a machine learning algorithm that produces a mathematical formula for calculating the next value in time given the previous values. Using this formula, we then attempt to re-create the known values and then we will calculate some more in order to predict the future. What we see here is that the model output (solid black line) for the time from zero to 6500 reproduces the known denoised data (dotted line) so well that we can hardly tell them apart on the diagram. That is a good sign.

Beyond reproducing the known data, the model is then used to compute values for the time from 6500 to 8000 minutes. On the image, we have also graphed the uncertainty in this prediction as we can no longer validate the prediction made due to a lack of observations in the future. And so we see a line that slowly gains a greater and greater uncertainty but remains within a well-defined corridor of values. The formula found, the model, thus reproduces the known data and makes a prediction that, on face value, makes sense. Whether this now corresponds to true reality, only experimentation (waiting for the future) can reveal.

Once we are confident that the model can indeed predict the future, then we can use the model to compute the future. On this basis, we may then, with confidence, plan actions to prevent events we do not want to occur but of which we know now that they will occur if we do nothing.

6.1.1 Data Preparation

We will assume that the data is already clean and so only contains data that is representative of the problem that we wish to model; see chapter 4 for methods of getting raw data this far. A factor that must be considered at this point is whether the data is in the form that allows a machine learning algorithm to quickly learn the salient features of the dataset.

We give a simple example to illustrate the point. Consider modeling the expected rise/fall of a stock price as a function of the principal balance sheet components of a company. To train this model, you may have data over several years from many companies. Such a model will not be useful to predict a share price (too little data per company) but it will reveal some interesting basic information. Two of the principal balance sheet figures are, for example, the stock price and the earnings. The machine learning algorithm will be able to learn that the expected rise/fall of the stock price depends upon the ratio price-to-earnings but it will require both time and data for making this conclusion. Since this is something that we humans already know, it would have helped the algorithm if we had removed the column of price and that of earnings and had inserted a column consisting of the ratio.

This example is referred to by the general injunction that one should add *domain knowledge* into the training data set. We have not attempted to explain the dynamics of the stock exchange to a neural network but rather we transformed the data into a form that is conducive for learning just as we would purchase a colorful nice language learning book for our child as opposed to a mere dictionary to aid it in its learning of a foreign language.

Another feature of data preparation for machine learning is that many learning methods do not deal well with data that is collinear. What is meant by this is that if we have two series of observations (e.g. a and b) that are related by a simple linear transformation (e.g. $a = mb + c$ with m and c constants), then the learner can become confused by this. The reason why this is strange may be compared to a situation with human beings. If we are trying to teach someone the meaning of "chair" and illustrate this with examples of the same chair in a small and large variety. The person who is to learn the concept may actually confuse the purpose of the chair with the size relationship in the dataset and thus learn something entirely unintended. This must be avoided by removing too simply related data from the dataset.

It cannot be overemphasized that the form in which data is presented to a machine learning algorithm has more influence on the accuracy of the final model than the choice of learning algorithm (as long as it is a reasonable algorithm that has the essential capability to learn the present problem). Thus, data preparation is a delicate activity and must be aided by thought and domain knowledge.

Generally speaking, the following questions should be answered when modeling industrial data:

1. Are all the relevant measurements present and have they been cleaned in the sense of chapter 4?
2. Are only the relevant measurements present?
3. Have simple relationships been removed?
4. Can the data be transformed in some meaningful way in order to better represent the phenomenon that is important for modeling?

6.1.2 How much data is enough?

When modeling, the question of how much data is needed always arises. Often, we are limited by practicality. Getting data sometimes represents a cost, perhaps both in time, money and effort. Having more data may also make learning slower. Increasing the sample rate may just produce noise and not more valuable signal information. Too little information is not enough and too much may be counterproductive as well. Before we end up with too much data, there is definitely a long interval of diminishing returns from getting more and more data.

One can philosophically say that an industrial process contains a certain amount of information regardless of how much data we extract from it. We must make sure that this information is represented in the data and by a reasonable amount of data. Thus, we must choose the right amount of data that represents the greatest gain of knowledge for the resources that we put into the acquisition of the data. Because this is dependent upon the problem at hand, we need a quantitative measure of "enough." We will approach this in two steps.

First, we will say that what we are really looking for is that if we have a reasonable model of the situation, then an additional bit of data will improve the model somewhat. As soon as we have reached the state where additional data does not allow a model improvement anymore, we may stop and say that we have enough data. Thus, we define enough data as that amount of data that we need to get the model to converge (to the right result), i.e. the model output for the validation data agrees with the experimental measurements for the same dataset and no longer changes appreciably with more data.

Second, we need to find a quantitative measure of convergence. This is provided, for example, by the variance. After the model is generated, we compute the variance, add more data, remodel and again compute the variance. In this way, we obtain the relationship between the size of the dataset and the variance. This relationship is roughly logarithmic, i.e. the variance rises with increasing data set size and eventually settles down to a more or less horizontal line. This can easily be detected and there is your convergence point.

Equally well, this can be measured by the mean squared deviation between model output and experimental data in the validation data set. This should behave in roughly the same manner and thus provide the point at which training may profitably stop.

Note that in this approach, it is not possible to calculate, *a priori*, how many data points are needed. Rather it is a checking procedure, in dependence upon the modeling method, to check if we already have enough. Any method to compute a definite data volume before modeling begins will be a mere estimate. Note also that the information content is not a simple function of data volume as, for example, many repetitions of the same measurement do not add any information. The data volume must, therefore have a suitably diverse population of points in it to aid the analysis.

6.2 Neural Networks

The term *neural network* refers to a wide family of functional forms that are frequently used to model experimental data. Historically, the creation of some of these forms was motivated by the apparent design of the human brain. This historical motivation aspect does not concern us here and will not be discussed.

For the present book, we shall distinguish between a *functional form* and a *function* by allowing the former to have numerically-undetermined parameters and require the later to specify numerical values for all parameters. So, for example, $\sin(ax)$ with a a parameter is a functional form and $\sin(2x)$ is a function. This distinction is crucial in machine learning as the principal effort in machine learning is always put into the methods of determining the numerical value of the parameters in an *a priori* determined functional form.

Practically speaking, we begin with a dataset and decide (by some unspecified and generally human procedure) that a certain functional form should be able to model that data. Then a machine learning algorithm determines the values of the parameters. The end result is a function (without parameters) that models the data.

The topic of neural networks can be profitably split into two categories

1. A list of different functional forms, so-called networks, that can be used to model data and their attendant properties such as

   a. The kind of functions that can be represented,
   b. Restrictions on the values of parameters,
   c. Robustness properties, and
   d. Scaling behavior.

2. A presentation of various algorithms used to determine the numerical value of the parameters in the functional form and their attendant properties such as

   a. Requirements on the training data form (labeled or unlabeled) or cleanliness (signal-to-noise-ratio or other pre-processing requirements),
   b. Speed of training,
   c. Convergence rate, convergence target (local or global optimum) and robustness of convergence, and
   d. Practical issues such as termination criteria, parametrization or initialization requirements.

Most books on neural networks mix these topics strongly and implicitly place a focus on the second. In practice, however, it is important to distinguish the network from the algorithm used to train it because we usually have a (largely) independent choice for both network and algorithm and must be aware of the involved limitations. For understanding, it is also important to know what a network can accomplish in the realistic case.

Training a neural network is a black art requiring much experience and depends mostly on the person preparing the data (pre-processing) and selecting the training methodology and parametrization of the training method. Even then, the issue of convergence to local optima typically requires significant tuning to a particular problem before a function is found that represents the data well enough for practical purposes[1]. For this reason, we will not be discussing training algorithms at all but refer to the specialist literature for this purpose [69]. We mention in passing that if you have a problem that you wish to model using neural networks and you are not already an expert in training them, it is probably best to get an expert to do the modeling for you as learning how to do it can require many months.

This chapter is intended to give the novice an overview of what neural networks are, what they can and cannot do and to give a sense of the complexity of the topic. Before going into the details, we need to discuss two important issues.

---

[1] Together, the neural network topology and the training method give rise to several variables having to be chosen by the human modeler. On top of that, most training methods involve some degree of random number generation, which means that each training run will be conducted slightly differently and so results cannot be completely comparable. The exact effect of changing any one of the initial parameters is unclear and so much stabbing in the dark may become necessary before enough learning has happened in the modeler's mind.

First, neural network methods yield a function that describes a specific set of data. What is first done to obtain this set of data or what is later done with this formula is no longer an aspect of neural network theory or practice. As such, it is useful to view a neural network as a summary of data – a large table of numbers is converted into a function – similar to the abstract of a scientific paper being a summary. Please note, that a summary cannot contain more information than the original set of data[2]; indeed it contains less! Due to its brevity, however, it is hoped that the summary may be useful. In this way, neural networks can be said to transform information into knowledge, albeit into knowledge that still requires interpretation to yield something practically usable.

Second, neural networks are intended for practical modeling purposes. The summarization of data is nice but it is not sufficient for most applications. To be practical, we require interpolative and extrapolative qualities of the model. Supposing that the dataset included measurements taken for the independent variable x taking the values $x = 1$ and $x = 2$, then the model has the *interpolative quality* if the model output at a point in between these two is a reasonable value, i.e. it is close to what would have been measured had this measurement been taken, e.g. at $x = 1.5$. Of course, this can only hold if the original data has been taken with sufficient resolution to cover all effects. The model has the *extrapolative quality* if the model output for values of the independent variable *outside* the observed range is also reasonable, e.g. for $x = 2.5$ in this case. If a model behaves well under both interpolation and extrapolation, it is said to *generalize* well.

A neural network model is generally used as a black-box model. That is, it is not usually taken as a function to be understood by human engineers but rather as a computational tool to save having to do many experiments and determine values by direct observation. It is this application that necessitates both above aspects: We require a function for computation and this is only useful if it can produce reasonable output for values of the independent variables that were not measured (i.e. compute the result of an experiment that was not actually done having confidence that this computed result corresponds to what would have been observed had the experiment been done).

The data used for training can be obtained from actual experiments or alternatively from simulations – neural networks are not concerned with the data source. Simulations of physical phenomena are often very complex as they are usually done from so-called first principles, i.e. by using the basic laws of physics and so on to model the system. As these simulations take time, they cannot be continuously run in an industrial setting. Neural networks provide a simpler empirical device by adjusting the parameters of a functional form until the so-determined function represents the data well.

---

[2] Whatever is in the data will hopefully also be in the network but there is no guarantee of this as the summarization process is a lossy process. Whatever is not in the data, however, is definitely not in the network. We must thus not expect a neural network to divine effects that have not been measured in the data. Issues such as noise and over-representation of one effect over another can produce unexpected results. That is the reason why pre-processing is so important.

This represents the major advantages of neural networks: They are (1) easier to create than first principles simulations, (2) can capture more dynamics than normal first principles model due to them modeling actual experimental output and not idealized situations, (3) once existing, are easy and fast to evaluate. They are thus practical and cheap. The price that must be paid for this practicality is that the way in which the function is obtained and the resultant function itself are both not intended for human understanding. Questions such as 'why' and 'how' must thus never be asked of a neural network. We may only ask 'what' the value of some variable is.

6.3 Basic Concepts of Neural Network Modeling

The dataset that will be used as the basis for obtaining the neural network contains several variables. For better understanding, we include a very simple example in table 6.1. We have three variables x_1, x_2 and y_1 in this table. Each variable has an associated measurement uncertainty or error $\Delta(x_1)$, $\Delta(x_2)$ and $\Delta(y_1)$. It is important to note that no observation whatsoever is fully accurate and so there are always measurement errors. Frequently, however, their presence is ignored for basic modeling purposes. We include them here because they have significant effects for industrial practice.

x_1	$\Delta(x_1)$	x_2	$\Delta(x_2)$	y_1	$\Delta(y_1)$
1.2	0.1	2.3	0.2	0.1	0.01
1.4	0.1	3.1	0.2	0.2	0.01
1.6	0.2	3.2	0.2	0.3	0.02
1.8	0.2	3.5	0.3	0.4	0.05

Table 6.1 An example data set for training a neural network. Note the presence of uncertainty measurements as well. This is important in practice as no measurement in the real world is totally precise.

The variables must first be classified into dependent and independent variables or, to use neural network vocabulary, into output and input variables respectively. We will decide that the two x variables are independent and yield the single y variable that is the dependent variable. We could have arbitrarily many independent and dependent variables in general; there is no fundamental limitation.

Thus, we look for some function $f(\cdots)$, such that $y_1 = f(x_1, x_2)$. In general, when we have many variables, we represent their collection by a vector, $\mathbf{x} = \{x_1, x_2, \cdots\}$. The general function is thus,

$$\mathbf{y} = f(\mathbf{x})$$

knowing Δx_i, we may compute Δy by

$$(\Delta \mathbf{y})^2 = \sum_i \left(\frac{\partial f(\mathbf{x})}{\partial x_i} \right)^2 (\Delta x_i)^2 .$$

Having classified the variables into input/output for the function, we must classify the type of output variable. There are two principal options. An output variable may be *numeric* or *nominal*. If a variable is numeric, its numerical value has some significance and so it makes sense to compare various value to each other, e.g. a temperature measurement. The function to be modeled is thus a regular mathematical function that could be drawn on a plot. If a variable is nominal, then the value of the variable serves only to distinguish it from other values and its numerical value has no significance. We use nominal values to differentiate category 1 from category 2 knowing that it is senseless to say that the difference between category 2 and category 1 is 1.

Neural networks are very often used for nominal variables and many are specifically intended to be *classification* networks, i.e. they classify an observation into one of several categories. Empirically, it has been found that neural network methods are very good at learning classifications.

Generally, most neural network methods assume that the data points illustrated in the above sample table are independent measurements. This is a pivotal point in modeling and bears some discussion.

Suppose we have a collection of digital images and we classify them into two groups: Those showing human faces and those showing something else. Neural networks can learn the classification into these two groups if the collection is large enough. The images are unrelated to each other; there is no cause-effect relationship between any two images – at least not one relevant to the task of learning to differentiate a human face from other images.

Suppose, however, that we are classifying winter versus summer images of nature and that our images were of the same location and arranged chronologically with relatively high cadence. Now the images are not independent but rather they have a cause-effect relationship ordered in time. This implies that the function $f(\cdots)$ that we were looking for is really quite different,

$$\mathbf{y} = f(\mathbf{x}) \qquad \rightarrow \qquad \mathbf{y_i} = f(\mathbf{x_{i-1}}, \mathbf{x_{i-2}}, \cdots, \mathbf{x_{i-h}}) .$$

In this version, we see a dependence upon history that implies a time-oriented memory of the system over a time length h that must somehow be determined. Depending on the dynamics of the time-dependent system, the memory of the process does not have to be a universal constant, so that in general $h = h(i)$ is itself a function of time.

As a consequence of this, we have network models that work well for datasets with independent data points (see section 6.4) and others that work well for datasets in which the data points are time-dependent (see section 6.5). The networks that deal with independent points are called feed-forward networks and form the historical beginning of the field of neural networks as well as the principal methods being used. The networks that deal with time-dependent points are called recurrent networks, which are newer and more complex to apply.

6.4 Feed-Forward Networks

The most popular neural network is called the *multi-layer perceptron* and takes the form

$$\mathbf{y} = \mathbf{a}_N \left(\mathbf{W}_N \cdot \mathbf{a}_{N-1} \left(\mathbf{W}_{N-1} \cdots \mathbf{a}_1 \left(\mathbf{W}_1 \cdot \mathbf{x} + \mathbf{b}_1 \right) \cdots + \mathbf{b}_{N-1} \right) + \mathbf{b}_N \right)$$

where N is the number of *layers*, the $\mathbf{W}_i$ are *weight matrices*, the $\mathbf{b}_i$ are *bias vectors*, the $a_i(\cdots)$ are *activation functions* and $\mathbf{x}$ are the inputs as before.

The weight matrices and the bias vectors are the place-holders for the model's parameters. The so-called *topology* of the network refers to the freedom that we have in choosing the size of the matrices and vectors, and also the number of layers N. Per layer, we thus get to choose one integer and thus have $N + 1$ integers to choose for the topology. The only restriction that we have is the problem-inherent size of the input and output vectors. Once the topology is chosen, then the model has a specific number of parameters that reside in these matrices and vectors.

The activation functions are almost always functions with the shape of a sinusoid, for example $\tanh(\cdots)$.

In training such a network, we must first choose the topology of the network and the nature of the activation functions. After that we must determine the value of the parameters inside the weight matrices and bias vectors. The first step is a matter of human choice and involves considerable experience. After decades of research into this topic, the initial topological choices of a neural network are still effectively a black art. There are many results to guide the practitioners of this art in their rituals but these go far beyond the scope of this book. The second step can be accomplished by standard training algorithms that we mentioned before and also will not treat in this book (see e.g. [69]).

A single-layer perceptron is thus

$$\mathbf{y} = \mathbf{a} \left(\mathbf{W} \cdot \mathbf{x} + \mathbf{b} \right)$$

and was one of the first neural networks to be investigated. The single-layer perceptron can represent only linearly separable patterns. It is possible to prove that a two-layer perceptron with a sigmoidal activation function for the first layer and a linear activation function for the second layer can approximate virtually any function of interest to any degree of accuracy if only the weight matrices and bias vectors in each layer are chosen large enough. In practice, we find that perceptrons with between two and four layers are used very frequently to model data.

Such a two-layer perceptron looks like

$$\mathbf{y} = m \left(\mathbf{W}_2 \cdot \tanh \left(\mathbf{W}_1 \cdot \mathbf{x} + \mathbf{b}_1 \right) + \mathbf{b}_2 \right)$$

where m is a scalar and where we get to choose the size of the two weight matrices to match the problem.

This approach is used both for numerical data and for nominal data and is found to work very well indeed. Many industrial applications are based on perceptrons, e.g. adaptive controllers.

6.5 Recurrent Networks

The basic idea of a recurrent neural network is to make the future dependent upon the past by a similar form of function like the perceptron model. So, for instance, a very simple recurrent model could be

$$\mathbf{x}(t+1) = \mathbf{a}\left(\mathbf{W}\cdot\mathbf{x}(t) + \mathbf{b}\right).$$

Note very carefully three important features of this equation in contrast to the perceptron discussed before: (1) Both input and output are no longer vectors of values but rather vectors of functions that depend on time, (2) there is no separate input and output but rather the input and output are the same entity at two different times and (3) as this is a time-dependent recurrence relation, we need an initial condition such as $\mathbf{x}(0) = \mathbf{p}$ for evaluation.

The above network, if we choose the activation function to be

$$a(z) = \begin{cases} 1 & z > 1 \\ z & -1 \le z \le 1 \\ -1 & z < -1 \end{cases},$$

is called the *Hopfield network* and is a very good classifier. The methodology is like this: (1) Every one of the possible categories is characterized by a vector of values called a *primary pattern*, (2) each item to be classified is also characterized by a similar vector of values, (3) the network is trained using a specialized training algorithm, (4) the characterizing vector of an as-yet-unclassified item is input into the network as the initial condition $\mathbf{p}$, (5) the network is now iterated in "time" until the vector converges to an unchanging state, (6) this convergent state is one of the primary patterns and thus the classification is done. This network uses the concept of time to accomplish a static task in which actual time does not play a role. If correctly constructed, the time iteration can make the network numerically more stable and so produce more reliable answers than for a static network like the perceptron.

Fig. 6.3 The first three digits as input to train a Hopfield network to recognize these digits.

We demonstrate the issues by means of a common example, recognition of digits. In figure 6.3 we display the bit pattern of three digits that we wish to recognize using a Hopfield network. Each two-dimensional pattern can easily be converted into a single dimensional vector of bits by joining each column on the bottom of the previous column. Thus, the digit 0" becomes the vector

$$\mathbf{x}_0 = [01111010000110000110000110000101110]^T .$$

With this setup, we can train the Hopfield network and obtain the matrix $\mathbf{W}$ and the vector $\mathbf{b}$. Using this network, we may then classify new inputs.

Figure 6.4 displays the results of verifying the network on novel input. We see that if we occlude 50% of the original pattern, we retrieve the correct result. However, if we occlude 67%, then we will get errors in recognition. If the network is presented with noisy inputs, the network will make an identification that is the same as a human being would have made. Thus, we conclude that the network is sensible in its classification.

We have thus been able to represent the difference between the first three digits in a Hopfield network. This is roughly the principle by which optical character recognition is done even though fancier techniques are being used in commercial software to make the system less error prone.

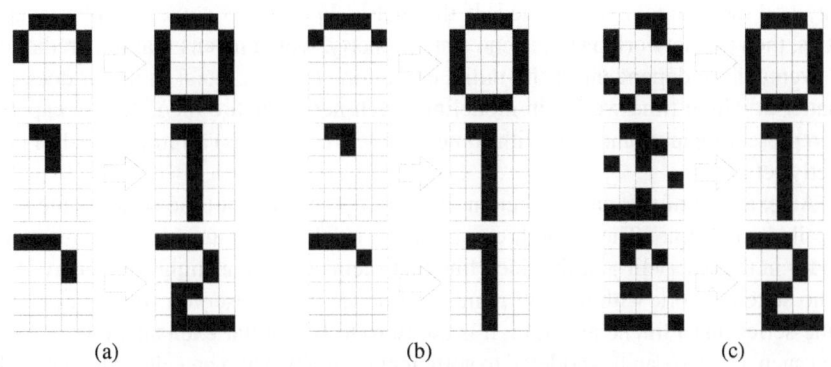

Fig. 6.4 Several test patterns for the digit recognizing Hopfield network and their associated outputs. Test set (a) includes 50% occluded inputs. Test set (b) includes 67% occluded inputs. Test set (c) includes noisy inputs.

A very new type of recurrent neural network is the *echo state network*. This uses the concept of a reservoir, which is essentially just a set of nodes that are connected to each other in some way. The connection between nodes is expressed by a weight matrix $\mathbf{W}$ that is initialized in some fashion[3]. The current state of each node in the reservoir is stored in a vector $\mathbf{x}(t)$ that depends on time t.

An input signal is given to the network $\mathbf{u}(t)$ that also depends upon time. This is the actual time-series measured in reality that we wish to predict. The input process is governed by an input weight matrix $\mathbf{W}^{in}$ that provides the input vector values to any desired neurons in the reservoir. The output of the reservoir is given as a vector $\mathbf{y}(t)$. The system state then evolves over time according to

$$\mathbf{x}(t+1) = f\left(\mathbf{W}\cdot\mathbf{x}(t) + \mathbf{W}^{in}\cdot\mathbf{u}(t+1) + \mathbf{W}^{fb}\cdot\mathbf{y}(t)\right)$$

where $\mathbf{W}^{fb}$ is a feedback matrix, which is optional for cases in which we want to include the feedback of output back into the system and $f(\cdots)$ is a sigmoidal function, usually $\tanh(\cdots)$.

The output $\mathbf{y}(t)$ is computed from the extended system state $\mathbf{z}(t) = [\mathbf{x}(t); \mathbf{u}(t)]$ by using

$$\mathbf{y}(t) = g\left(\mathbf{W}^{out}\cdot\mathbf{z}(t)\right)$$

where $\mathbf{W}^{out}$ is an output weight matrix and $g(\cdots)$ is a sigmoidal activation function, e.g. $\tanh(\cdots)$.

The input and feedback matrices are part of the problem specification and must therefore be provided by the user. The internal weight matrix is initialized in some way and then remains untouched. If the matrix $\mathbf{W}$ satisfies some complex conditions, then the network has the echo state property, which means that the prediction is eventually independent of the initial condition. This is crucial in that it does not matter at which time we begin modeling. Such networks are theoretically capable of representing any function (with some technical conditions) arbitrarily well if correctly set up.

An example of this can be seen in figure 6.5. The original time-series is the very detailed spiky line. The smooth curve on top is an echo state network with many nodes in the reservoir and the thick line that seems to have a slight time delay is an echo state network with a small number of nodes in the reservoir. In this case, the time-series has a financial origin, it is the Euro to US Dollar exchange rate. We see that such a signal can be modeled to sufficient accuracy using an echo state network.

Please note again at this point the principal difference between a time-series in which the points are correlated in time and the classification of observations into categories in which the observations are not correlated at all. The correlation in time makes it necessary to use much more sophisticated mathematics.

---

[3] Generally it is initialized randomly but substantial gain can be got when it is initialized with some structure. At present, it is a black art to determine what structure this should be as it definitely depends upon the problem to be solved.

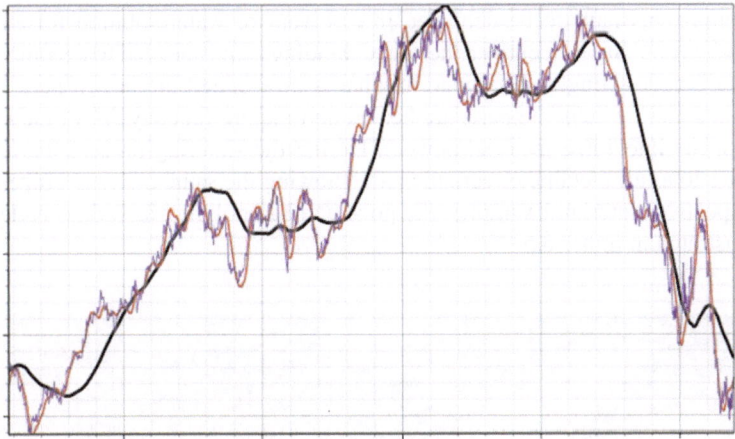

Fig. 6.5 The prediction of the Euro to US Dollar exchange rate over 1.5 years.

6.6 Case Study: Scrap Detection in Injection Molding Manufacturing

Co-Authors: Pablo Cajaraville, Reiner Microtek
Björn Dormann, Klöckner Desma Schuhmaschinen GmbH
Dr. Philipp Imgrund, Fraunhofer Institute IFAM
Maik Köhler, Klöckner Desma Schuhmaschinen GmbH
Lutz Kramer, Fraunhofer Institute IFAM
Oscar Lopez, MIM TECH ALFA, S.L.
Kaline Pagnan Furlan, Fraunhofer Institute IFAM
Pedro Rodriguez, MIM TECH ALFA, S.L.
Dr. Natalie Salk, PolyMIM GmbH
Jörg Volkert, Fraunhofer Institute IFAM

The injection molding technology is a widely used technology for the mass production of components with complex geometries. Almost all material classes can be processed with this technology. For polymers the pelletized material is injection molded under elevated temperature in a mold cavity showing the negative structure of the resulting part. The part is cooled down and ejected to the finished component.

In the case of metals and ceramics the so called metal injection molding (MIM) or ceramic injection molding (CIM) process is applied. Both processes fall under the umbrella term powder injection molding (PIM). In all cases, the material powder is mixed with a binder system composed of polymers and/or waxes. This so-called feedstock is subsequently injection molded compared to the described polymer material. The ejected parts are called green parts still contain the binder material that acted as flowing agent during the injection molding process. To remove the binder, the components have to be debinded in a solvent or water solution for a certain amount of time. Subsequently, a thermal debinding step is needed to decompose the residual binder acting as backbone in what is now called the brown part. During the final sintering step, the parts are heated up to approximately 3/4 of the melting point of the integrated material powder. During the sintering process the material densifies to a full metallic or ceramic part, showing the same material properties as the respective material. Examples of a good and bad polymer as well as metal parts are illustrated in figure 6.6.

Fig. 6.6 Both images display a good part and a damaged part. On the left, we have a plastic part where we can have various deformations as one damage mechanism. On the right, we have a green metal part that is broken in one place and a final metal part showing the end product as it should be. In both examples, the damage is visible but this is often not so.

Often, the damages to a part that occur during injection can only been seen on the final part. The unnecessarily performed steps of debindering and sintering have expended significant amounts of electrical energy and have also made the material useless. If we could identify a damaged part during its green stage, we could recycle the material and also save the energy for debindering and sintering. For all parts, there is an effort involved in determining whether the part is good or not. Today, this effort is usually made manually, which is expensive. If we could make the identification automatic, then we would save this effort as well.

An injection molding machine is controlled by manually inputting a series of values known as set-points. These are the values for various physical quantities that we desire to have during the injection process. It is the responsibility of the machine to attempt to realize these set-points in actual operation. This attempt is generally achieved but there are deviations in the details. In order to monitor what the actual value of these various measurements is, an injection machine will also have sensors that output these measurements over time, i.e. a time-series.

For each part produced, we thus have the set-points and also a variety of time-series over the duration of its injection. This information is available in order to characterize a part.

We will assume that the function that computes the scrap versus good status, the decision function, takes the form of a three-layer perceptron

$$\gamma = \mathbf{W}_3 \cdot \tanh(\mathbf{W}_2 \cdot \tanh(\mathbf{W}_1 \cdot \mathbf{x} + \mathbf{b}_1) + \mathbf{b}_2) + \mathbf{b}_3$$

where the bias vectors $\mathbf{b}_i$ and the weight matrices $\mathbf{W}_i$ must be determined by some training algorithm.

In order to make an input vector $\mathbf{x}$, we must extract salient features from the time-series in the form of scalar quantities that allow the characterization of scrap vs. good parts as we cannot input the whole time-series into the neural network. If we did input the entire time-series, we would force the weight matrices to become extremely large. Thus we would have many parameters to be found by training. This would require many more parts to be produced and this is unrealistic. We must live with a few hundred training parts and so we must keep the input vector as small as possible. After many trials with various settings, we have found the best performance with the following procedure (see figure 6.7 for an example):

1. Take all observations of good parts over the pilot series. For each time series, create an averaged time series over all these good parts.
2. Disregarding local noise in the time-series, compute the turning points of this time series.
3. For every part encountered, perform the same turning point analysis.
4. For every part, we now compute the difference between the turning points in the present part relative to the turning points of the averaged series. If we find differences, then these will be taken as salient features.
5. These differences form the salient features and these will constitute the vector $\mathbf{x}$. For practicality, we limit ourselves to a specific maximum number of turning points allowed.

In order to use the training method, it must be provided with some pairs $(\mathbf{x}, \gamma)$ of input vectors and the quality output. To determine these pairs, we must manually assess a number of injected parts and characterize them as scrap or good. Having obtained such training data, the training algorithm produces the decision function. In practise this means that the machine user must inject several parts, record the data $\mathbf{x}$, manually determine whether the final parts are scrap or not, γ, and provide this information to the learning method. We have determined that a good number of observations is 500 or more. Furthermore, it is good to use several settings of the set-points within these 500 parts.

When a new part is now injected, its data $\mathbf{x}$ is provided to the decision function and it computes whether this part is good or bad γ and outputs this value. As a result a robot can be triggered to remove the scrap parts from further production.

We may define four different recognition rates:

1. *good rate*, ρ_g – The number of correctly identified good parts
2. *scrap rate*, ρ_s – The number of correctly identified scrap parts
3. *false-negative rate*, ρ_n – The number of good parts identified as scrap
4. *false-positive rate*, ρ_p – The number of scrap parts identified as good

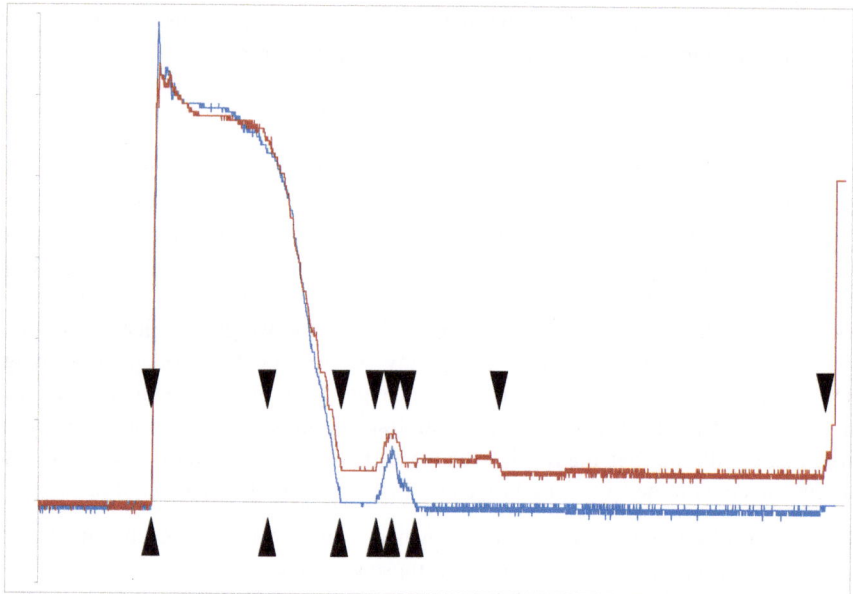

Fig. 6.7 The bottom curve is the average pressure observed at the nozzle averaged over all parts known to be good. The top curve is a single observation of a part known to be bad. The black arrows pointing up indicate the position of the turning points of the bottom curve and the black arrows pointing down indicate the position of the turning points of the top curve. We observe that the top curve has two extra turning points. We also observe that the vertical position of several turning points is higher than that of the average good curve.

where each count is divided by the total number of produced parts in order to make each item into a genuine rate. Please note that these rates are thus normalized by definition, i.e. $\rho_g + \rho_s + \rho_n + \rho_p = 1$. It is generally not possible to design a system that will have perfect recognition efficiency ($\rho_n = \rho_p = 0$). We would rather throw away a good part than let a scrap part through. Thus, the overall objective is to minimize the false-positive rate ρ_p by designing a decision function that is as accurate as possible.

This enumeration is, of course, theoretical as it would require the user to know which parts are *really* good or bad. This identification would require the manual characterization that we want to avoid using the present methods (except for the training data set for which it is necessary). Thus, we will never actually know what these rates are except in two cases: the pilot series where the data is used for training the function and, possibly, in any quality control spot checks that are usually too infrequent to really allow the computation of a rate. Thus, these rates must be interpreted as a useful guideline for thinking but not practical numerical quantities except at training time when we actually know the quality state of all parts.

Even if we are able to correctly identify every part as either good or bad, this will not change the amount of scrap actually produced – it will merely change the amount of scrap delivered to the customer. In order to reduce the amount of scrap actually produced, we must interact with the production process actively via the set-points and this is our next stage.

We observe that the recognition rates mentioned above are in fact functions of the set-points. In particular, we want to reduce the scrap rate as we would like to produce as many good parts as possible. Let us combine the ten set-points α_i into a single vector,

$$\mathbf{s} = (\alpha_1, \alpha_2, \cdots, \alpha_{10}). \tag{6.1}$$

Using this, we thus focus on the scrap rate function $\rho_s(\mathbf{s})$. This function achieves a global minimum ρ_s^* at the point $\mathbf{s}^*$,

$$\rho_s^* = \min_{\mathbf{s}}\{\rho_s(\mathbf{s})\} = \rho_s(\mathbf{s}^*). \tag{6.2}$$

The point $\mathbf{s}^*$ is determined using an optimization algorithm and then communicated to the injection molding machine.

These methods were tried on the part in the right image of figure 6.6. In total 500 parts were made, with various settings of the set-points, of which approximately 20% were scrap. This high scrap rate results from the various set-points settings, some of which are, of course, not optimal.

We find that the recognition efficiency is 98% in that $\rho_g + \rho_s = 0.98$ where we recognize good and bad parts correctly nearly always with 490 out of 500 parts. We obtain a low false positive rate $\rho_p = 0.002$ in which scrap parts are recognized as good. Relative to our sample size of 500, this means that a single scrap part was not recognized as such. The false negative rate in which good parts are recognized as scrap was $\rho_n = 0.018$, which means 9 parts in total.

Recall that the objective of training the network was to minimize the false positive rate. It is clear that this cannot be reliably zero and so getting a rate of one part in 500 can be interpreted as a success. The system is certainly more reliable than manual quality control, which is common in the industry.

The actual production scrap rate was approximately 20%. This could have been reduced to 5% by adjusting the set-points appropriately. Of course, a real production will not be at the level of 20% scrap and so an improvement of a factor of four seems unlikely. Nevertheless, a significant factor should be possible.

Now that we have verified that it is possible to reliably identify scrap from good parts based only on process data and that we can optimize the process based on the same analysis, we ask what the practical significance of this is. There are two major points: Quality improvement and energy savings.

With respect to quality improvement, there are also two aspects.

First, the quality of the delivered product. Even if we produce 20% scrap, 98% of these are correctly recognized to be scrap and so these are not delivered to the client. Looking at the current numbers, then, we produce 500 parts all in all. Of these, we have 100 scrap parts. We recognize 98 scrap parts as scrap, 392 good parts as good, 9 good parts as scrap and one scrap part as good. Thus the client receives 392 good parts and one scrap part in the delivery. This is an effective scrap rate – with the client – of 0.3%. The identification was thus able to lower the production scrap rate of 20% to a delivery scrap rate of 0.3%.

Second, the production quality will also improve due to the optimization. Since we can lower the production scrap rate of 20% to 5%, we would produce 475 good parts and 25 scrap parts as compared to the above figures. In final consequence, the client would receive 466 parts of which one would be scrap. This lowers the effective delivery scrap rate to 0.2% and relative to a larger delivery size. With the optimization, the molding production cost per delivered part is lowered by 19%. This is a reduction in production cost that is otherwise unreachable.

With respect to energy savings, we also save the energy costs that would have flown into the steps of debinding and sintering of parts that later are recognized as scrap. It is hard to quantify this in any general manner but we assume that this lowers the total production cost per delivered part by another 4%.

We gratefully acknowledge the partial funding of this research by the European Union: Investment in your future – European fund for regional development, the EU programs MANUNET and ERANET, the German ministry for education and research (BMBF) and the Wirtschaftsförderung Bremen (economic development agency of the city state of Bremen, Germany).

6.7 Case Study: Prediction of Turbine Failure

In this study we will attempt to predict a known turbine failure using historical data for it. We refer to section 5.5 for details on turbines and coal power plants. On a particular turbine, a blade tore off and completely damaged the turbine. After the event, the question was raised whether this could have been predicted and localized to a specific location in the turbine.

The specific turbine in question has over 80 measurements on it that were considered worthwhile to monitor. Most of these were vibrations but there were also some temperatures, pressures and electrical values. A history of six months was deemed long enough and the frequency depended upon each individual measurement point – some were measured several times per second, others only once every few hours. In fact, the data historian only stores a new value in its database if the new value differs from the last stored value by more than a static parameter. In this way, the history matrix contained a realistic picture of an actual turbine instrumented as it normally is in the industry. No enhancements were made to the turbine, its instrumentation or the data itself. A data dump of six months was made without modification.

The data stopped two days before a known (historically occurring) blade tear on that turbine. During the time leading up to the blade tear and until immediately before it, no sign of it could be detected by any analysis run by the plant engineers either before or after the blade tear was known. Thus, it was concluded that the tear is a spontaneous and thus unpredictable event.

Initially, the machine learning algorithm (echo state network from section 6.5) was provided with no data. Then the points measured were presented to the algorithm one by one, starting with the first measured point. Slowly, the model learned more and more about the system and the quality of its predictions improved both absolutely and in terms of the maximum possible future period of prediction. Once even the last measured point was presented to the algorithm, it produced a predication valid for the following two days of real time. The result may be seen in figure 6.8.

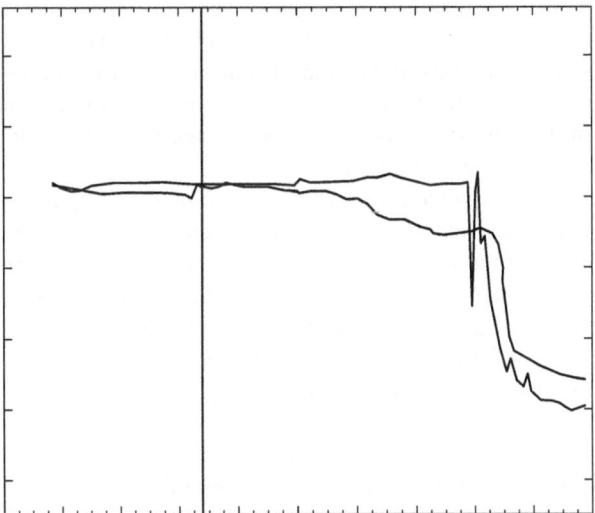

Fig. 6.8 Here we see the actual measurement (spiky curve) versus the model output (smooth line) over a little history (left of the vertical line) and for the future three days (right of the vertical line). We observe a close correspondence between the measurement and the model. Particularly the event, the sharp drop, is correctly predicted two days in advance.

Thus, we can predict accurately that something will take place in two days from now with an accuracy of a few hours. Indeed it is apparent from the data that it would have been impossible to predict this particular event more than two days ahead of time due to the qualitative change in the system (the failure mode) occurring only a few days before the event. The model must be able to see some qualitative change for some period of time before it is capable of extrapolating a failure and so the model has a reaction time. Events that are caused quickly are thus predicted relatively close to the deadline but two days warning is enough to prevent the major damages in this case. In general, failure modes that are slower can be predicted longer in advance.

It must be emphasized here that the model can only predict an event, such as the drop of a measurement. It cannot label this event with the words "blade tear." The identification of an event as a certain type of event is altogether another matter. It is possible via the same sort of methods but would require many examples of blade tears and this is a practical difficulty. Thus, the model is capable of giving a specific time when the turbine will suffer a major defect; the nature of the defect must be discovered by manual search on the physical turbine.

This is interesting but to be truly helpful, we must be able to locate the damage within the large structure of the turbine, so that maintenance personnel will not spend days looking for the proverbial needle in the haystack.

Fault detection and localization is now done by performing an advanced data-mining methodology (singular spectrum analysis from section 4.5.2) that tracks frequency distributions of signals over the history and can deduce qualitative changes. Over the 80 measurements points, we are able to isolate that four of them contain a qualitative shift in their history and that two of these four go through such a shift several days before the other two. Thus, we are able to determine which two out of 80 locations in the turbine are the root cause for the event that is to occur in two days. See figure 6.9 for an illustration.

In this figure, we graph the abnormality as measured by singular spectrum analysis (see section 4.5.2) over time for each measurement. We observe that only four time-series are abnormal at all. Two of them become abnormal early in time and two others follow. When we ask which time-series these are, then we find that the first two are the radial and axial vibrations of one bearing and the second two are the same vibrations of the neighboring bearing. We may safely attribute causation to this evolution. Thus we say that the first bearing's abnormality causes the second bearing's abnormality. We also say that the first bearing's abnormality was the first sign of what later lead to the event, the blade tear. Indeed, the blade tore off very close to this particular bearing. Thus we are successful in localizing the fault within the large turbine.

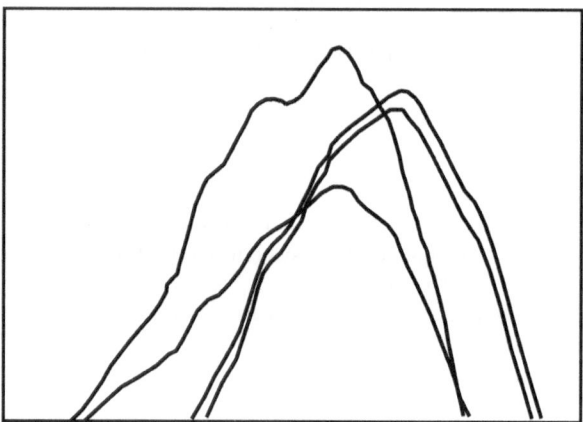

Fig. 6.9 We compute a deviation from normal being tracked over a window of about four days length. So we observe that two sensors start behaving abnormally and two days later, two other sensors behave abnormally. About 3.5 days after the start of the abnormal behavior, this new behavior has become normal and so the deviation from normal is seen to reduce again. Therefore, we observe a qualitative change in the performance of these four points.

The localization that is possible here is to identify the sensor that measures an abnormal signal and that will be the first to show the anomaly that will develop into the event. It is, of course, not possible to compute a physical location on the actual turbine more accurately than the data provided. However, a physical search of the turbine, after the actual blade tear, found out that the cause was indeed at the location determined by the data-mining approach.

It is possible to reliably and accurately predict a failure on a steam turbine two days in advance. Furthermore, it is possible to locate the cause of this within the turbine so that the location covered by the sensor that measures the anomaly can be focused on by the maintenance personnel. The combination of these two results, allows preventative maintenance on a turbine to be performed in a real industrial setting saving the operator a great expense.

6.8 Case Study: Failures of Wind Power Plants

Wind power plants sometime shut down due to diverse failure mechanisms and must be maintained. Especially in the offshore sector but also in the countryside, are these maintenance activities costly due to logistics and delay. Common failures are for example due to insufficient lubrication or bearing damages. These can be seen in vibration patterns if the signal is analyzed appropriately. It is possible to model dynamic evolving mechanisms of aging in mathematical form such that a reliable prediction of a future failure can be computed. For example, we can say that a bearing will fail in 59 hours from now because the vibration will then exceed the allowed limits. This information allows a maintenance activity to be planned in advance and thus saves collateral damage and a longer outage.

Wind power plants experience failures that lead to financial losses due to a variety of causes. Please see figure 6.10 for an overview of the causes, figure 6.11 for their effects and figure 6.12 for the implemented maintenance measures. Figure 6.13 shows the mean time between failures, figure 6.14 the failure rate per age and figure 6.15 the shutdown duration and failure frequency[4]. From these statistics we may conclude the following:

1. At least 62.9% of all failure causes are internal engineering related failure modes while the remainder are due to external effects, mostly weather related.
2. At least 69.5% of all failure consequences lead to less or no power being produced while the remainder leads to ageing in some form.
3. About 82.5% of all maintenance activity is hardware related and thus means that a maintenance crew must travel to the plant in order to fix the problem. This is particularly problematic when the power plant is offshore.
4. On average, a failure will occur once per year for plants with less than 500 kW, twice per year for plants between 500 and 999 kW and 3.5 times per year for

---

[4] All statistics used in figures 6.10 to 6.15 were obtained from ISET and IWET.

plants with more than 1 MW of power output. The more power producing capacity a plant has, the more often it will fail.

5. The age of a plant does not lead to a significantly higher failure rate.
6. The more rare the failure mode, the longer the resulting shutdown.
7. A failure will, on average, lead to a shutdown lasting about 6 days.

From this evidence, we must conclude that internal causes are responsible for a 1% capacity loss for plants with less than 500 kW, 2% for plants between 500 and 999 kW and 3.5% for plants with more than 1 MW of power output.

In a wind power field like Alpha Ventus in the North Sea, with 60 MW installed and expecting 220 GWh (i.e. an expectation that the field will operate 41.8% of the time) of electricity production per year, the 3.5% loss indicates a loss of 7.7 GWh. At the rate of German government regulation of 7.6 Eurocents per kWh, this loss is worth 0.6 million Euro per year. Every cause leads to some damage that usually leads to collateral damages as well. Adding the cost of the maintenance measures related to these collateral damages themselves yields a financial damage of well over 1 million Euro per year. The actual original cause exists and cannot be prevented but if it could be identified in advance, then these costs could be saved.

This calculation does not take into account worst case scenarios such as the plant burning up thus requiring effectively a new build.

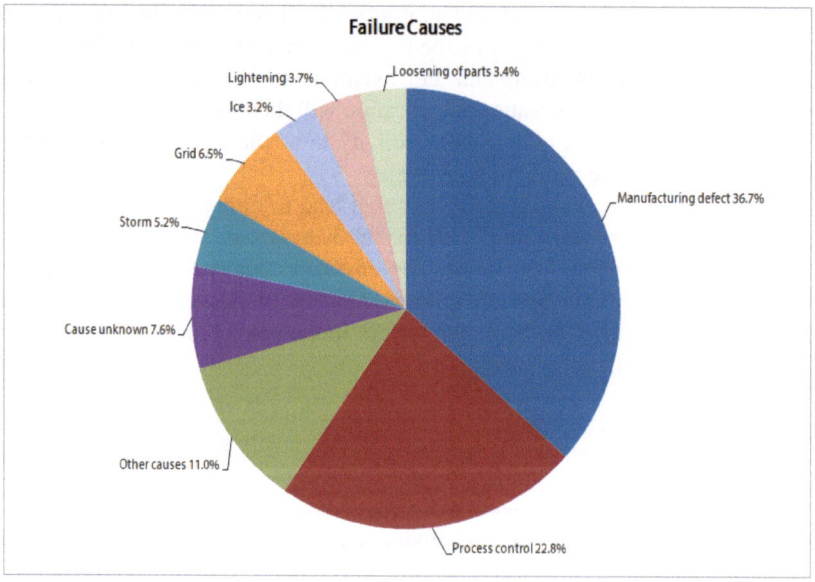

Fig. 6.10 The causes for a wind power plant to fail are illustrated here with their corresponding likelihood of happening relative to each other.

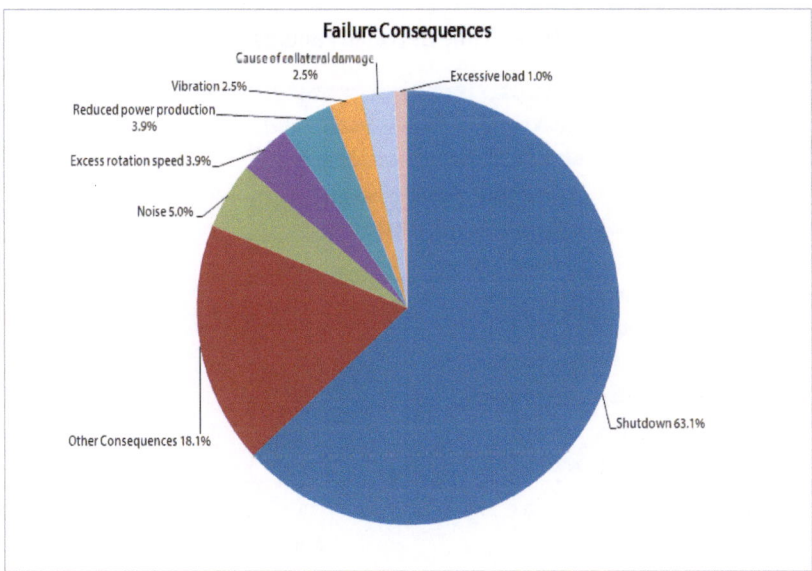

Fig. 6.11 The effects of the causes of figure 6.10 are presented here with the likelihood of happening relative to each other.

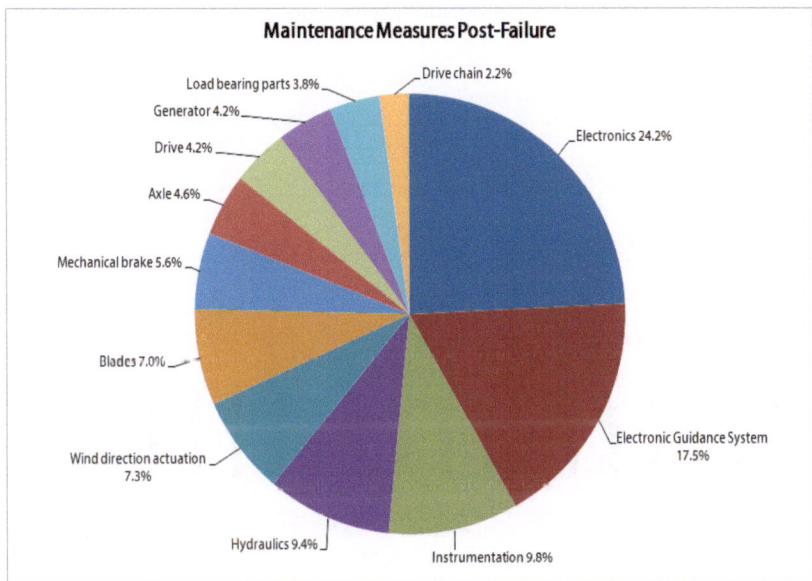

Fig. 6.12 The maintenance measures put into place to remedy the effects of figure 6.11 with the likelihood of being implemented relative to each other.

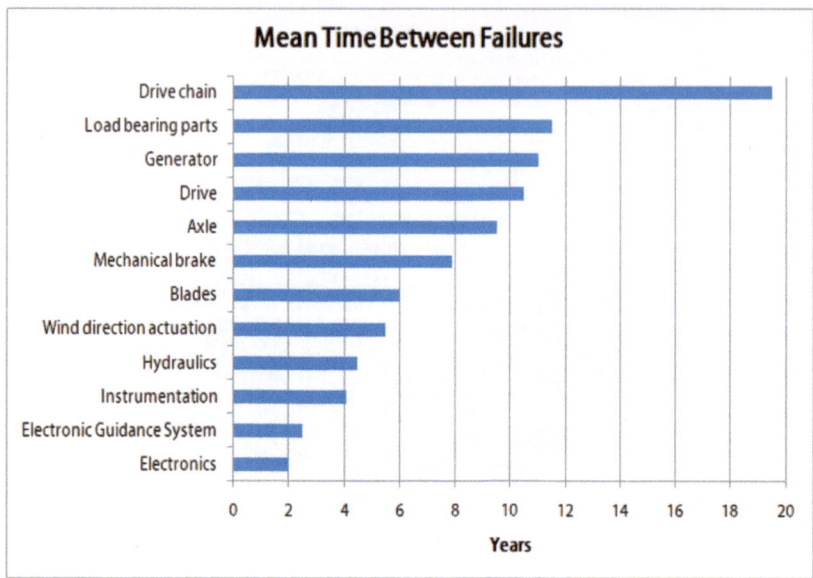

Fig. 6.13 The mean time between failures per major failure mode.

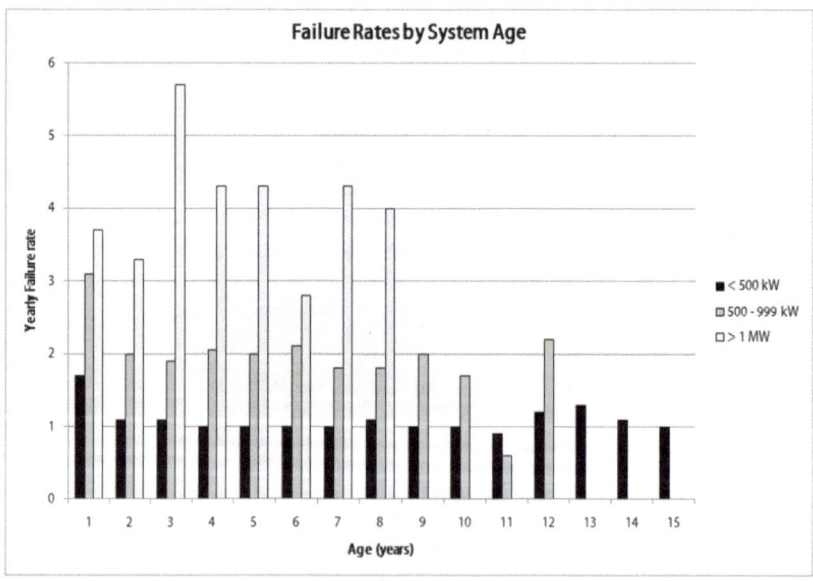

Fig. 6.14 The yearly failure rate as a function of wind power plant age. It can be seen that plants with higher output fail more often and that age does not significantly influence the failure rate.

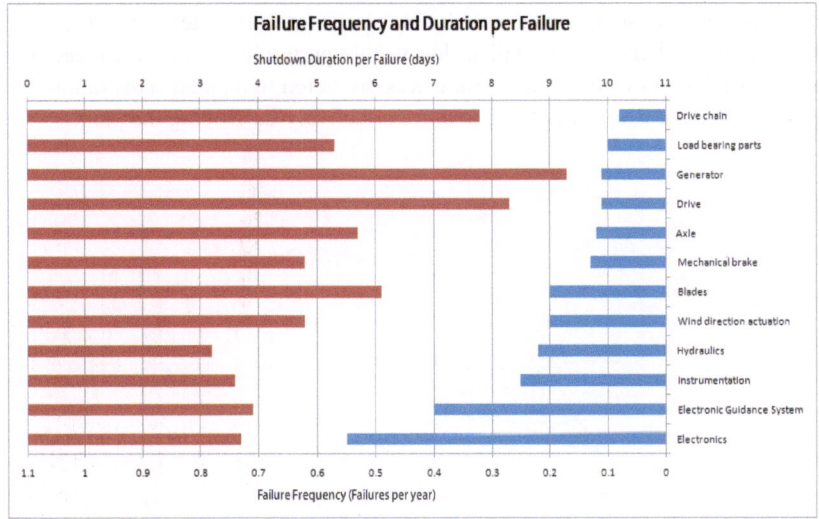

Fig. 6.15 The failure frequency per failure mode and the corresponding duration of the shutdown in days.

A recurrent neural network was applied to a particular wind power plant. From the instrumentation, all values were recorded to a data archive for six months. One value per second was taken and recorded if it different significantly from the previously recorded value. There were a total of 56 measurements available from around the turbine and generator but also subsidiary systems such a lubrication pump and so on. Using five months of these time-series, a model was created and found that the model agreed with the last month of experimental data to within 0.1%. Thus, we can assume that the model correctly represents the dynamics of the wind power plant.

This system was then allowed to make predictions for the future state of the plant. The prediction, according to the model's own calculations, was accurate up to one week in advance. Naturally such predictions assume that the conditions present do not change significantly during this projection. If they do, then a new prediction is immediately made. Thus, if for example a storm suddenly arises, the prediction must be adjusted.

One prediction made is show in figure 6.16, where we can see that a particular vibration on the turbine is to exceed the maximum allowed alarm limit after 59 ± 5 hours from the present moment. Please note that this prediction actually means that the failure event will take place somewhere in the time range from 54 to 64 hours from now. A narrower range will become available as the event comes closer in time. This information is however, accurate enough to become actionable. We may schedule a maintenance activity in two days from now that will definitely prevent the problem. Planning for two days in advance is sufficiently practical that this would solve the problem in practice.

In this case, no maintenance activity was performed in order to test the system. It was found that the turbine failed due to this particular vibration exceeding the limit after 62 hours from the moment it was predicted to happen. This failure led to contact with the casing, which led to a fire effectively destroying the plant.

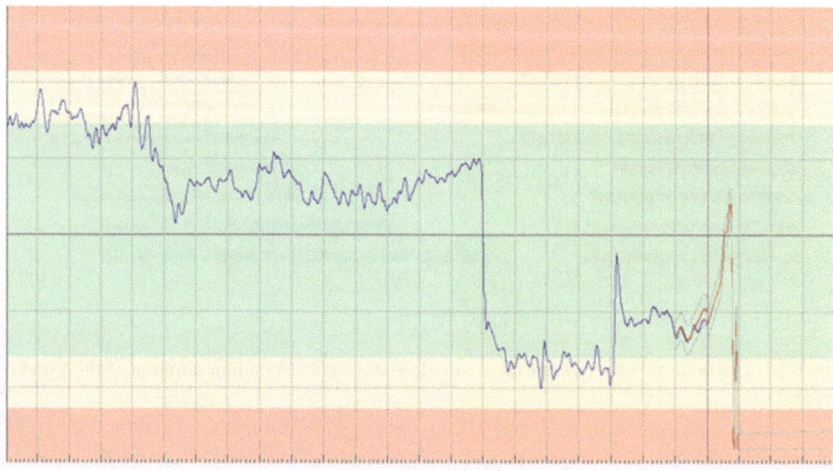

Fig. 6.16 The prediction for one of the wind power plant's vibration sensors on the turbine clearly indicating a failure due to excessive vibration. The vertical line on the last fifth of the image is the current moment. The curve to the left of this is the actual measurement, the curve to the right shows the model's output with the confidence of the model.

It would have been impossible to predict this particular event more than 59 hours ahead of time due to the qualitative change in the system (the failure mode) occurring just a few days before the event. The model must be able to see some qualitative change for some period of time before it is capable of extrapolating a failure and so the model has a reaction time. Events that are caused quickly are thus predicted relatively close to the deadline. In general, failure modes that are slower can be predicted longer in advance.

6.9 Case Study: Catalytic Reactors in Chemistry and Petrochemistry

Catalytic reactors are devices in chemical plants whose job it is to provide a conducive environment for a certain chemical reaction to take place, see figure 6.17. In a reactor, at least two substances are brought into contact with each other. One is a substance that we would like to change in some chemical way and the other is the catalyst, i.e. the substance that is supposed to bring this change about. The two substances are mixed and heated to provide the energy for the change. Then we wait and provide the necessary plumbing for the substances to come into and for the end product to leave the reactor. Some parts that are not converted have to be re-cycled back for a second, and possibly more, times through the reactor until finally all the original substance has been changed. An example is the breaking down of the long molecular chains of crude oil in the effort to make gasoline.

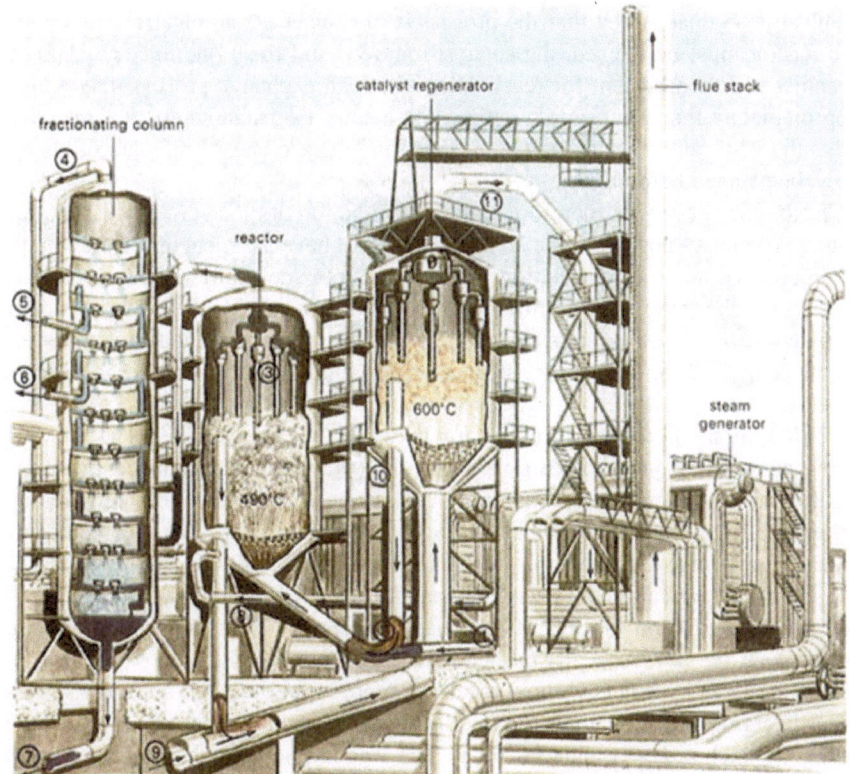

Fig. 6.17 The basic workings of the catalytic reactor system in a petrochemical refinery.

The catalyst performs its work upon the substance and brings about a change. It thereby uses up its potential to cause this change and thus ages over time. This degradation of the catalyst is the primary problem with operating such a reactor continuously over the long term. The catalyst must therefore be re-activated in some fashion and at some time.

We will investigate both of the major two kinds of catalytic reactors that exist: the *fluid catalytic converter* (FCC) and the *granular catalytic reactor* (GCR).

In the FCC, the catalyst is a fluid that can be pumped into and out of the reactor. We can therefore create a loop in which the catalyst is pumped into the reactor to perform its function and then out again into a reactivation phase only to return. This loop occurs forever and the catalyst can thus be used essentially without limit. However the speed of the loop must be carefully tuned to the actual aging of the catalyst inside the reactor so that we do not put either too much work (attempting to reactivate catalyst that is still fine) or too little work (not reactivating enough catalyst so that eventually we have too little) into the reactivation job.

In the GCR, the catalyst is in the form of granules that are filled into tubes in the reactor. These granules stay inside the tubes until the point is reached where the catalyst is so deactivated that the process is no longer economical. At this point, the reactor must be opened, the catalyst removed and fresh one injected. The old granules can then be sent for reactivation. Such an exchange process may require approximately four weeks of downtime and is thus a substantial cost for the plant. Also the new catalyst must be ordered well in advance and so the date of change must be planned beforehand.

Both types of reactor therefore require a prediction of the aging process into the future. We must know weeks in advance if we will have a critical deactivation.

In order to make the prediction, we have access to several temperatures around the reactor, the inflow and outflow, a gas chromatographic identification of what is flowing out and a few process pressures. In fact the age of the catalyst is measured by the pressure difference across the reactor. The higher it gets, the older the catalyst is.

Using the method of recurrent neural networks, we create a model of the GCR using almost four years of data in which the catalyst was exchanged twice, see figure 6.18. The jagged curve running over the whole plot is the measured pressure difference over the reactor. Whenever you see a sharp drop, this means that the catalyst has been exchanged; this happened three times in total in that figure.

The mathematical model draws the smooth curve over the data. We can hardly see the difference on the left of the image, so closely does the model represent the data. At the first vertically dashed line, we have reached the "now" point from which the model predicts without receiving more input data. We see three smooth lines diverging from this time. The middle one is the actual prediction, the other two being the uncertainty boundaries for it. The jagged line then shows what actually occurred. We can see that the model is very accurate indeed. The brief ups and downs in the real measurement are not in fact due to aging of the catalyst but due to various operational modes and varying quality of the crude oil being injected in the reactor. We are only concerned with the long term trend and not with short term fluctuations.

At the time of the second vertically dashed line, the catalyst was exchanged. The prediction was accurate up to this time, 416 ± 25 days later. Thus the prediction was accurate over one year in advance.

You may ask why the catalyst was exchanged, for the three times that we see on the figure, at different ages (or different differential pressures). Would it not be good to specify a single boundary value to define "too old?" In this case, our true cut-off criterion is not a certain age but rather the point of uneconomicality of the process. As this depends on various scientific but also on various economic influences, the

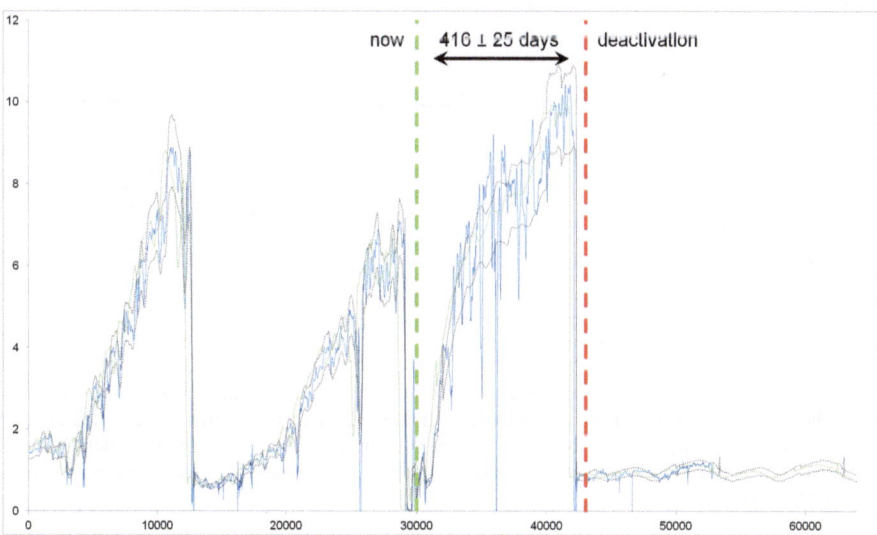

Fig. 6.18 The pressure differential (equivalent to the catalyst age) over time. The jagged line is the actual measurement and the middle smooth line the prediction. The first vertical line from the left to the right is the present moment from which the prediction starts. The second vertical line is the time at which we predict a catalyst exchange to be necessary. This prediction can be seen, using the real measurements, to be correct more than one year in advance.

actual age at which the catalyst becomes uneconomical changes with fluctuating market prices. These (and their uncertainties) must be taken into account.

We now proceed to the FCC in a chemical plant. Here we will take a different viewpoint. The FCC is a complex unit that has many set-points. For instance the rate and manner in which the fluid catalyst is recycled is up to us to control. The set-points that control this are changed by the operating personell to match the demands of the time.

While measurements such as the ambient air temperature are variables that must be measured to be known, the set-points are of a different nature. As the operators modify the set-points in dependence upon market factors, we do have some knowledge about these beforehand. Thus, we ask to what extent does this prior knowledge help the model to predict the future state?

We investigate, in figure 6.19, a very simple neural network model (perceptron, see section 6.4) for the pressure differential in dependence upon all other variables of the FCC, of which there many be several dozen and several set-points as well. If we only provide historical information to it, we obtain the solid curve. Compared to the actually measured dotted curve, we can see that this model is too simplistic to be able capture reality. If we present the future data for the set-points in addition (i.e. the production plan) to the historical data to the same simple model, we obtain the dashed curve. This dashed curve is accurate and achieves our aim.

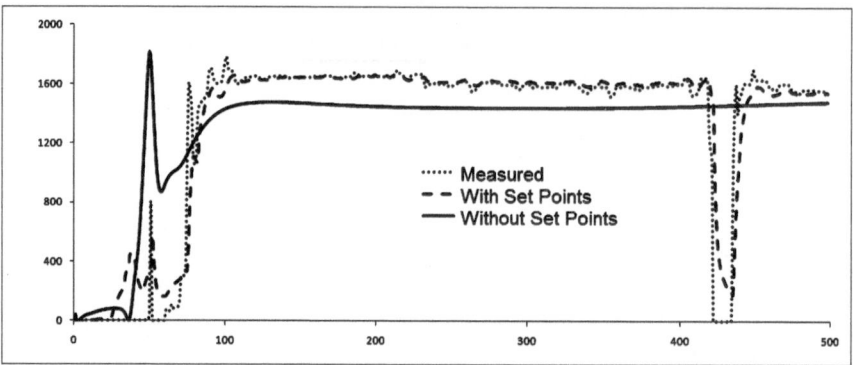

Fig. 6.19 The pressure differential over time (in hours) on a fluid catalytic converter predicted into the future using two different models. The actual measurement is shown in the dotted line. A prediction without any information about the future is shown in the solid line. A prediction made with the knowledge of some future set-points is shown in the dashed line. It is clear that knowledge of future actions is beneficial.

We can thus see that, in the case of a simple perceptron model, the provision of some limited information about the future significantly helps the model to predict those measurements that we cannot know in advance.

In conclusion, we see that both major kinds of catalytic reactors can be modeled well. We can predict both kinds of reactor into the future and thus provide information about essential future events such as the deactivation of the catalyst and thus the time (in the case of GCR) and the manner (in the case of FCC) of this deactivation. From both predictions, we easily derive the ability to plan specific actions to remedy the problem.

6.10 Case Study: Predicting Vibration Crises in Nuclear Power Plants

Co-Author: Roger Chevalier, EDF SA, R&D Division

So far, we have focused on predicting failure events. Such events are characterized by large, usually fast, changes that result in damage and usually a shutdown of the plant. In this section, we will focus on predicting a more subtle phenomenon. We observe that the vibration measurement on a certain bearing of a steam turbine increases periodically. This increase is alarming but does not represent a damage or danger.

In our specific example, the turbine has five bearings and each has a vibration sensor. See figure 6.20 for a plot of a temporary increase in vibration, which we will call vibration crisis. The exact cause of the problem is not precisely identified at present but it always in the same conditions of vacuum pressure and power

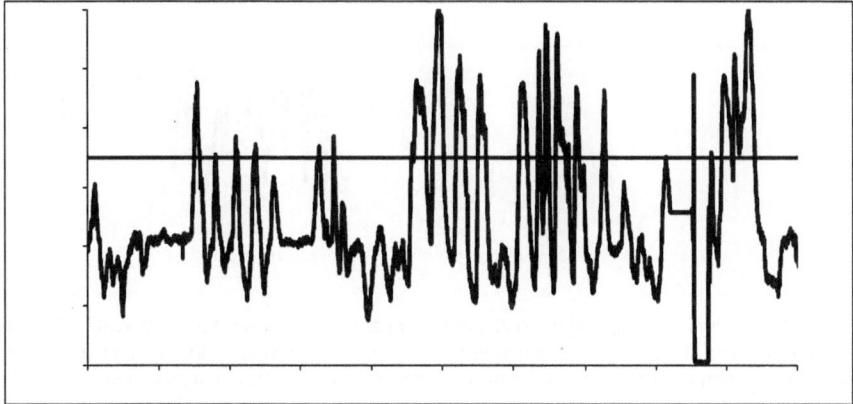

Fig. 6.20 This is the vibration of one bearing over time. The horizontal line is the limit for the vibration crisis, i.e. if the vibration measurement exceeds this limit, we speak of a crisis. It will be our goal to detect such events. Time is measured in units of ten minutes and so the plot is over a period of roughly 35 days.

This study concerns itself with the prediction of future vibration crises and not with determining the mechanism at its source. If one can know, only hours in advance, that a crisis will happen, this will help operators significantly in preparing for the event. The plant can be regulated into a state more conducive to controlling the impending crisis.

We attempt a model via a recurrent neural network (see section 6.5). The information about the turbine includes five displacement vibrations, for each bearing we have two metal pad temperatures, the steam pressure at several points in the process, the axial position of the turbine shaft, the rotation rate, the active and reactive power produced and one temperature on the oil circuit. This information is sampled once every ten minutes for the period of about five months in order to generate the model and learn the signs of an impending vibration crisis. The dataset contained several examples of such crises so that effective learning was possible.

With respect to the raw data from figure 6.20, we see the results of the prediction process in figure 6.21. The raw data is displayed in gray and the model output in black. We have a first period during which we observe the turbine before a prediction is possible. When enough data is there, we enter into a second period during which we predict and immediately validate the predictions with the real data. We see during this second period a close agreement between model output and measured values. Then we enter the third period, the actual future during which no more measurement data are available. This is the genuine prediction.

We note from figure 6.20 that we can correctly predict a vibration crisis, in this example, up to about 2.8 days in advance. Beyond this point, the prediction is no

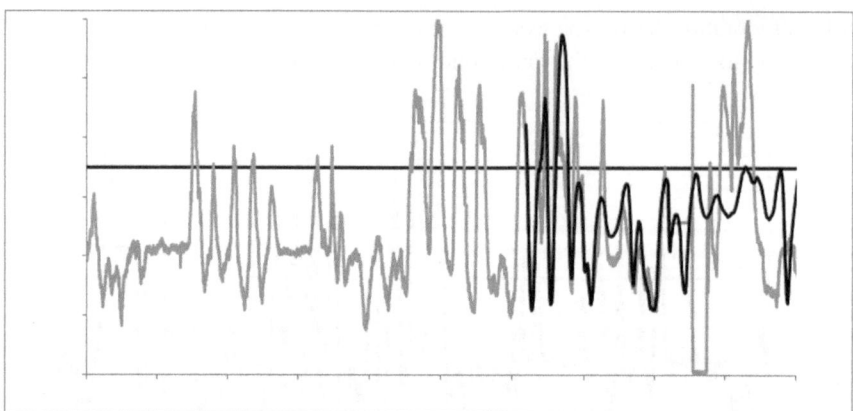

Fig. 6.21 The same data as in figure 6.20 but now in gray and overlayed in black with the model output from time 3100. We note that the vibration crisis from 3100 to 3500 is predicted correctly but the next vibration crisis at 3700 and the one after that from 4500 are not predicted correctly.

longer successful. This is roughly played out in all other examples. Thus, we see that there is a lead time of less than 3 days before such an event is detectable. This must be enough in practice to construct some kind of reaction.

Please note carefully the aim of this study. It is not the aim to correctly represent the vibration measurement over all values and all times. The goal is to accurately compute the times at which the vibration measurement crosses the limit line. When doing modeling it is essential to keep one's objectives clear as modeling is an adjustment of numerical parameters with respect to minimizing some sort of numeric accuracy requirement. In general, if one's aim were a representation of the vibration signal as such, one would choose the least squares method to measure the difference between measurement and model output. In this case, however, we are not trying to do this. Thus, our metric is not the least squares metric but the deviation between the actual and modeled times of the vibration signal crossing the limit line – a very different goal. Thus, the figure 6.20 should be interpreted accordingly.

We note that the strongest correlant with the problem is the outside temperature (cooling water). However it is not as clean as having a specific trigger temperature. The problem has a compound trigger in which the cooling water temperature plays a leading role but not the only role. A prediction of a future vibration crisis is reliable for 3 days in advance. If we attempt to predict it further into the future, the uncertainty in the prediction makes the prediction itself useless. Of course, the closer we get to the interesting time, the more accurate the prediction gets.

We have made 6 such predictions in a double blind study and have correctly predicted 5 vibration crises from among the 6 cases. The model is thus quite successful in being able to predict the future occurrence of a crisis. This is the case even though the specific causal mechanism remains still under investigation.. The model could be improved if the problem would be better understood so that the data can be more properly prepared..

6.11 Case Study: Identifying and Predicting the Failure of Valves

A chemical plant has a particular unit that is meant to combine several chemicals from a variety of input sources and to provide a gaseous output with an as-constant-as-possible composition. This task is controlled by an assembly of 40 valves that are controlled by a computer that opens and closes them according to a well-balanced schedule. If the valves fail to open and close according to schedule and are either fast or slow or leak, then the tailgas is not constant and causes problems later on in the process. In this study, we are to predict future problems of the assembly and also identify which of the 40 valves is responsible for the problem.

In the whole process, there are three phases. For each of these, we will compute the probability distribution of deviations between the set point provided by the control computer and the actual response as measured by the instrumentation. Figure 6.22 displays the results for each phase. We observe that one phase has an exponential distribution and the two others do not as they have secondary or even tertiary local maxima. An exponential distribution is what we would expect to see from normal operations of a controller – deviations are very rare and exponentially decreasing in magnitude indicating that deviations are random in origin. Seeing secondary peaks in this distribution is not expected as it shows a structured mechanism and hence some form of damage.

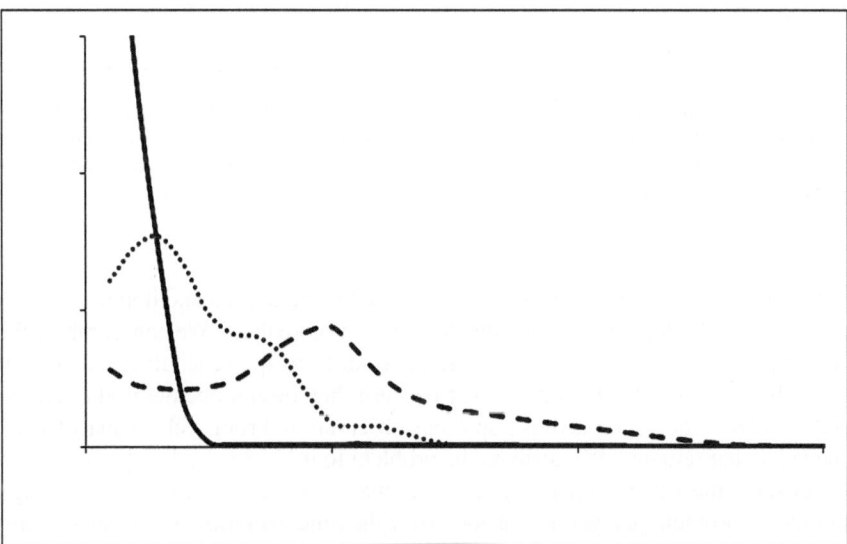

Fig. 6.22 The probability distribution over the three phases of valve operations. The vertical axis is the probability and the horizontal axis the normalized absolute value of the deviation between set point and actual response of valve openings. We observe that one phase appears to be operating correctly (exponential distribution) and two phases incorrectly (non-exponential distributions).

Next, we will introduce measure of abnormality for a valve. The score itself is based on the difference between set point and response (just as in figure 6.22). However, we also demand that there be an associated surge in non-constancy of the tailgas within a certain response time to track only those abnormal valve openings/closings that were close to a unwanted product surge. In figure 6.23, we graph the abnormality in this sense for each valve across all three phases of operation.

The valve numbers are on the horizontal axis and the absolute value of the difference between set point and measurement on the valve opening and closing is on the vertical axis. The solid, dashed and dotted lines correspond to the three phases in the same manner as in figure 6.22. The grey line is the average of the three phase lines.

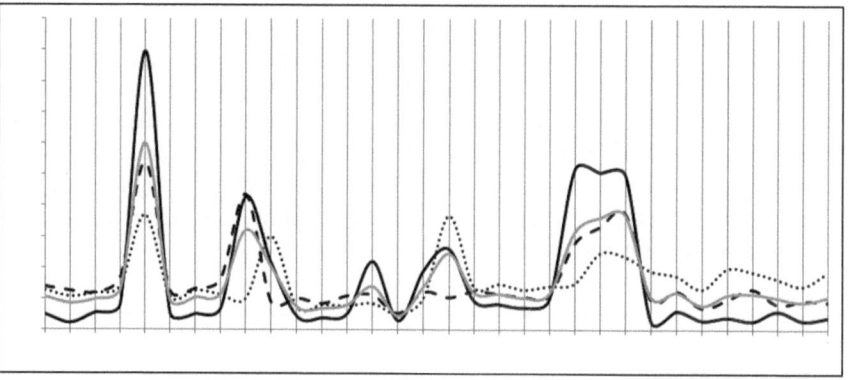

Fig. 6.23 A selection of the 40 valves (horizontal axis) is investigated here in terms of their deviation from the set point (vertical axis) in those cases in which an output pressure surge occurs within a certain time frame. The three black lines correspond to the three phases in figure 6.22 and the gray line is the average of the three black lines.

Combining these two images, we must look for the dashed and dotted phases in figure 6.23 and select those valves that have a high score there. We now combine this information with some process specific analysis from the scheduling application. From this we can rule out some valves because they do not operate in the relevant phase and their deviation is thus a spurious observation. From such an analysis, it is one valve that remains. We attribute the problem to it.

As such, the valve is operating in a way that is not ideal but it is not causing a significant problem just yet. Let us look over the time evolution of this abnormality in figure 6.24. The jagged line is the abnormality at each moment in time. As we are not concerned with the shorter term ups and downs, we take a 20-day moving average to smooth out the curve. This is the thick line on the plot. We observe an upward trend over time with a dip near the end. The time on the horizontal axis is in days and so this evolution occurs over a significant time frame. We now create a prediction of this time-series, which is plotted as the continuation of the thick line on the right of the image.

Please note that the peak observed in figure 6.24 at day 147 is a failure of a valve. After the valve has been fixed, we see the abnormality decline, which suggests that the maintenance measure has relieved the problem. However the abnormality does not decline to its former levels. This means that we have not solved the problem fully. From the prediction made, we predict that another failure is going to occur on day 208. This prediction is made on day 175, i.e. 33 days in advance. The uncertainty of this prediction is ±10 days.

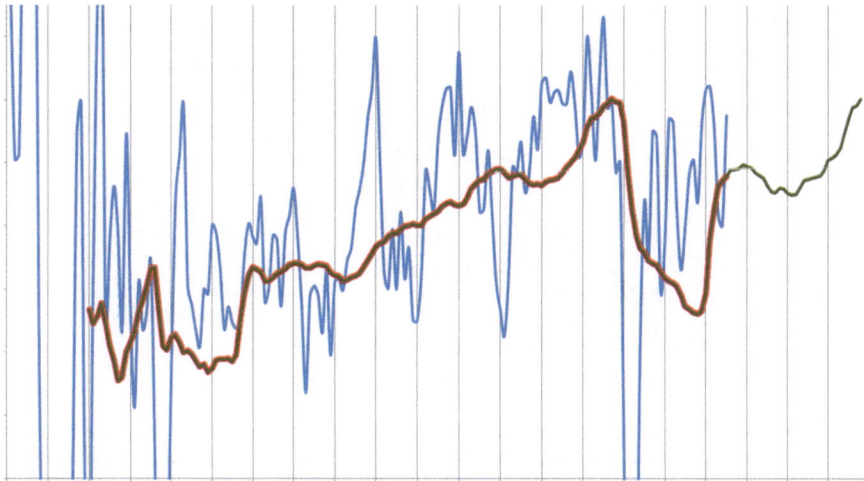

Fig. 6.24 Abnormality over time during the relevant phase together with 20-day moving average and prediction into the future. The peak around day 147 is a known failure. On day 175, we predict a second failure to occur on day 208 with an accuracy of ±10 days. This event occurred as predicted.

After waiting for a few weeks, we find that the failure did indeed happen as predicted. The failed valve is the same valve as the one we have identified using the above abnormality approach.

Thus we conclude that is possible to predict the future failure of valves and to identify which valve it will be even if we only have information about the family of valves.

6.12 Case Study: Predicting the Dynamometer Card of a Rod Pump

Co-Authors: Prof. Chaodong Tan, China University of Petroleum
          Guisheng Li, Plant No. 5 of Petrochina Dagang Oilfield Company
          Yingjun Qu, Plant No. 6 of Petrochina Changqing Oilfield Company
          Xuefeng Yan, Beijing Yadan Petroleum Technology Co., Ltd.

A rod pump is a simple device that is used the world over to pump for oil on land, see figures 6.25 and 6.26. Basically, we drill a hole in the ground and appropriately cement the hole such that a nice vertical cavity results. This cavity is filled with a rod that is going to move up and down using a mechanical device that is called the rod pump. Attached to the bottom of the rod is a plunger, which is a cylindrical "bottle" used to transport the oil. On the downward stroke, the plunger is allowed to fill with oil and on the upward stroke this oil is transported to the top where it is extracted and put into barrels.

Fig. 6.25 A schematic of a rod pump. The motor drives the gearbox, which causes the beam to tilt. This drives the horsehead up and down. This assembly assures that the rotating motion of the motor is converted into a linear up-down movement of the rod. The stuffing box contains the oil that is discharged through a valve on the top of the well.

Let us focus our attention on two variables of this assembly: the displacement of the rod as measured from its topmost position and the tension force in the rod. When we graph these two variables against each other such that the displacement goes on the horizontal and the tension on the vertical axis, we will find that, as the system is in cyclic motion, the curve is a closed locus. This is called the *dynamometer card* of the rod pump, see figure 6.27 (01) for an example of expected operations. To travel once around the locus takes the same amount of time as it takes the rod pump to complete one full cycle of downstroke and upstroke. A normal rod pump makes four strokes per minute.

It is a remarkable observation that the shape of this locus allows us to diagnose any important problem with the rod pump [64]. In figure 6.27 we display dynamometer card examples for the most common problems.

We will go into a little detail on these shapes and their problems because it is an exceptional fact that a complete diagnosis can be made so readily from a single image. This approach should be possible for a variety of other machines once only the correct measurements and the correct way of presenting them are found. That is the deeper reason behind presenting these here. It should be encouraged to seek a similar presentation of faults in other machinery.

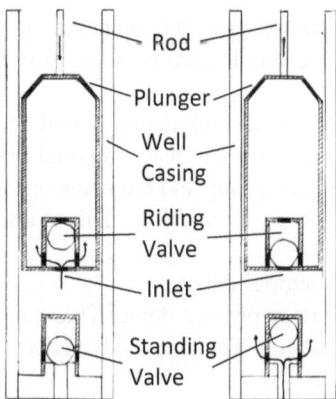

Fig. 6.26 A schematic of the well bottom. The rod drives the plunger down into the well guided by the well casing. The bottom of the plunger has a so-called riding valve to take in the oil through the inlets. The bottom of the well is closed off to the reservoir with a so-called standing valve that open once the plunger is at the bottom.

The cases are:

01 This is the shape we expect to see on a properly working rod pump. The upper and lower horizontal features are nearly parallel and the diagram is close to the theoretically expected diagram.

02 Another example of good operations

03 The two horizontal features are sloping downward, are much closer to each other and more wave-shaped than in the good case. This is due to excessive vibration of the rod.

04 The lower right-hand corner of the card is missing but the two horizontal features are still horizontal. This indicates that the plunger is not being filled fully but that the pump is working properly.

05 A more severe case of the former kind.

06 Here the pump is still working properly but the oil is very thick.

07 These distinctive jagged features with the lower right corner missing are caused by the presence of sand in the oil. This will cause damage to the rod assembly in the short term.

08 The lower right-hand corner is missing but the horizontal features are no longer horizontal; the bite taken out of the low right corner has an exponential boundary. This is caused by reservoir de-gassing and slowing the downward plunge.

09 A more severe case of the former kind.

10 A similar case to the former kind. Here the gas forms an air-lock inside the plunger preventing the plunger from draining at the top.

11 The bottom horizontal feature is rounded and/or lifted up making the whole card significantly smaller. This is due to a leaking inlet valve.

12 The opposite feature to above. Here the top horizontal feature is rounded and/or pressed down making the whole card smaller. This is due to a leaking outlet valve.

13 This oval feature results from a combination of both the inlet and the outlet valve leaking. Note that this is a fairly flat oval compared to the oval of image (06).

14 The top left-hand corner is missing and the boundary is in the shape of an exponential curve; compare with (08) and (09). This is due to a delay in the closing of the inlet valve.

15 Same as above but for a shorter delay.

16 The right side of the card is pressed down. This happens because of a sudden unloading of the oil at the top. The outlet valve is not opening properly but suddenly.

17 The characteristic upturned top right-hand corner (as opposed to (08)) indicates a collision of the plunger and the guide ring.

18 The lower left-hand corner is bent backwards and the top right-hand corner is sloped down in addition to features like (08). This indicates a collision between the plunger the fixed valve at the bottom of the hole.

19 The thin card with concave loading and unloading dents on upper left and lower right corners indicates a resistance to the flow of the oil such as the presence of paraffin wax.

20 Very thin but long card in the middle of the theoretically expected card with wavy horizontal features indicates a broken rod.

21 A thin long card with straight horizontal features indicates that the plunger is filling too fast due to a high pressure inside the reservoir. The plunger should be exchanged for a larger one.

22 The card looks normal but is too thin, particularly on the bottom. This is due to tubing leakage.

23 The piston is sticking to the walls of the hole and bending the rod.

It can easily be seen that the diagnosis of problems is immediate from the shape given a little experience in the matter. In fact, it has been shown that the diagnosis can be automated by recognizing the shape with a perceptron neural network [43] (see section 6.4 for a discussion of perceptrons).

Our purpose here is to investigate whether we can predict the future shape of the dynamometer card and thus diagnose a situation today that will lead to a problem in the next days.

In order to predict the evolution of the shape over time, we must be able to characterize the shape numerically first. For this we will seek a two step process. First, we will attempt to find a formula-based model for the shape itself that has only a few parameters that must be fitted to any particular shape. As we get a new dynamometer card several times per minute, this fitting process happens continuously thereby inducing a time-series on those parameters. It is these parameters that we will model using a recurrent neural network. In total, this will yield a prediction system for the future shape of a dynamometer card.

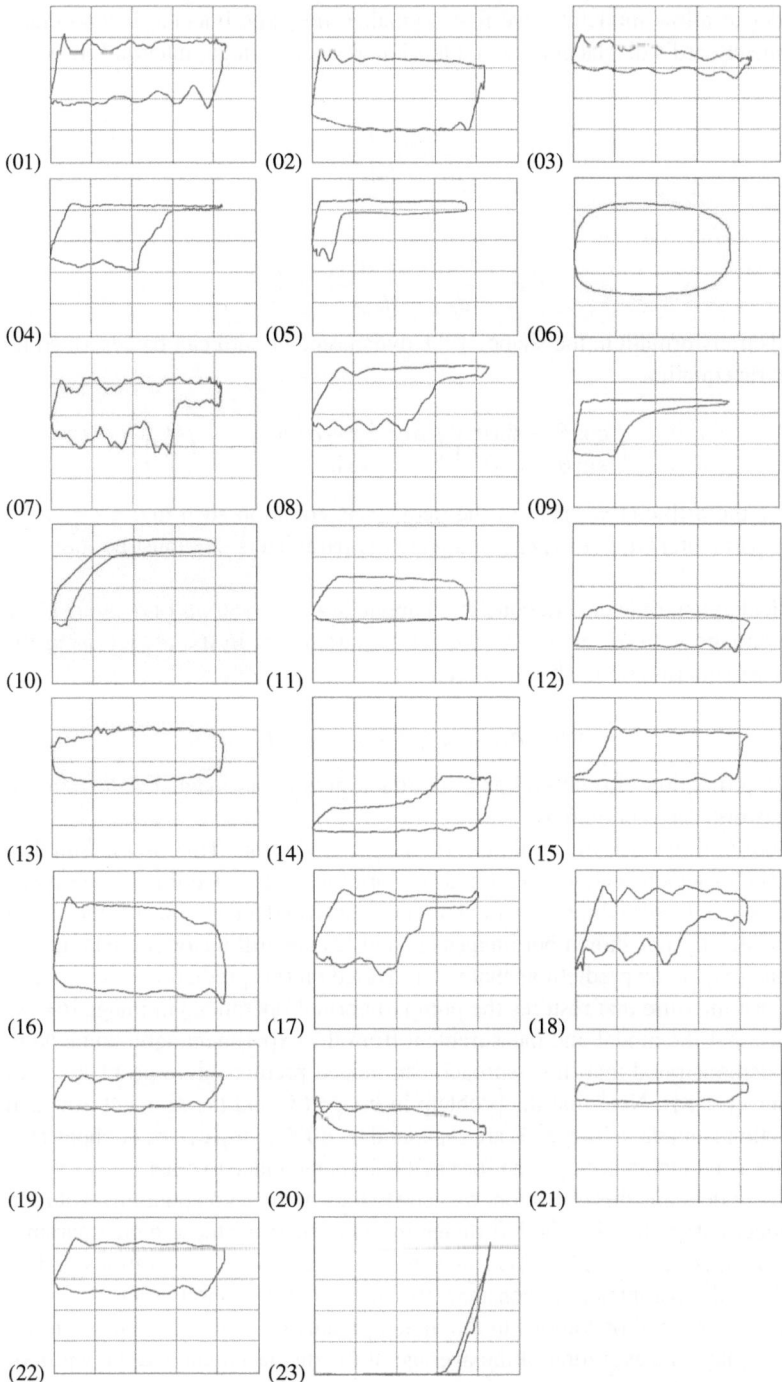

Fig. 6.27 The various cases of dynamometer cards. See text for an explanation.

In order to allow modeling, we first normalize the experimental data so that a dynamometer card's displacements and tension always lie in the interval [-1,1]. We do this by

$$t' = \frac{2t - t_{min} - t_{max}}{t_{max} - t_{min}}$$

and

$$d' = \frac{2d - d_{min} - d_{max}}{d_{max} - d_{min}}.$$

Using this transformation, the shape of the dynamometer card can be described by a parametric equation

$$\begin{pmatrix} t' \\ d' \end{pmatrix} = \begin{pmatrix} \cos\theta & -\sin\theta \\ \sin\theta & \cos\theta \end{pmatrix} \begin{pmatrix} a\cos^b x + c\sin^d x \\ e\sin x \end{pmatrix} + \begin{pmatrix} f \\ g \end{pmatrix}$$

where x is the artificial variable of the parametric equation such that $x \in [-\pi, \pi]$ [83]. The parameters a, b, c, d, e, f and g are the parameters that must be found by fitting.

A typical dynamometer card consists of about 144 observations of d and t. Thus, we have enough data to reliably fit the 7 free parameters in the model using the normal least-squares approach. The vector

$$\mathbf{v} = [a, b, c, d, e, f, g, t_{min}, t_{max}, d_{min}, d_{max}]$$

is actually a function of time $\mathbf{v}(t)$ and this function is then modeled as a recurrent neural network, see section 6.5.

In figure 6.28 we see the evolution of such a prediction. Time is measured in strokes each of which is about 15 seconds in duration. The dotted line is the experimental data and the solid line is the model. The historical data upon which the model is based is mostly not shown but images (1) and (2) are still historical data. Image (3), (4) and (5) are the predictions that result. Based on this prediction, we initiate a maintenance measure that restores the pump to normal operations in image (6). As we can see that the model and measurement from the experiment agree quite well, we have demonstrated that this approach can indeed predict future problems with dynamometer cards. Note that the problem in image (5) and the moment at which the prediction is made in image (3) are separated by 4000 pump cycles or about 16.7 hours. This is enough warning time for practical maintenance to react.

To fully understand this evolution, we need to look at the corresponding evolution of the model parameter. See the left image in figure 6.29 for those model parameters that changed. From time 15000 onwards, we have normal operations and so this level for the parameters is constant. Based on this, we observe an increasing deviation from normal operations in the model parameters. The right image in figure 6.29 displays the evolution of the average of the displacement and the width of the displacement. We display these as the experimental data was normalized for the images in figure 6.28. Here we also observe an abnormal behavior in the beginning.

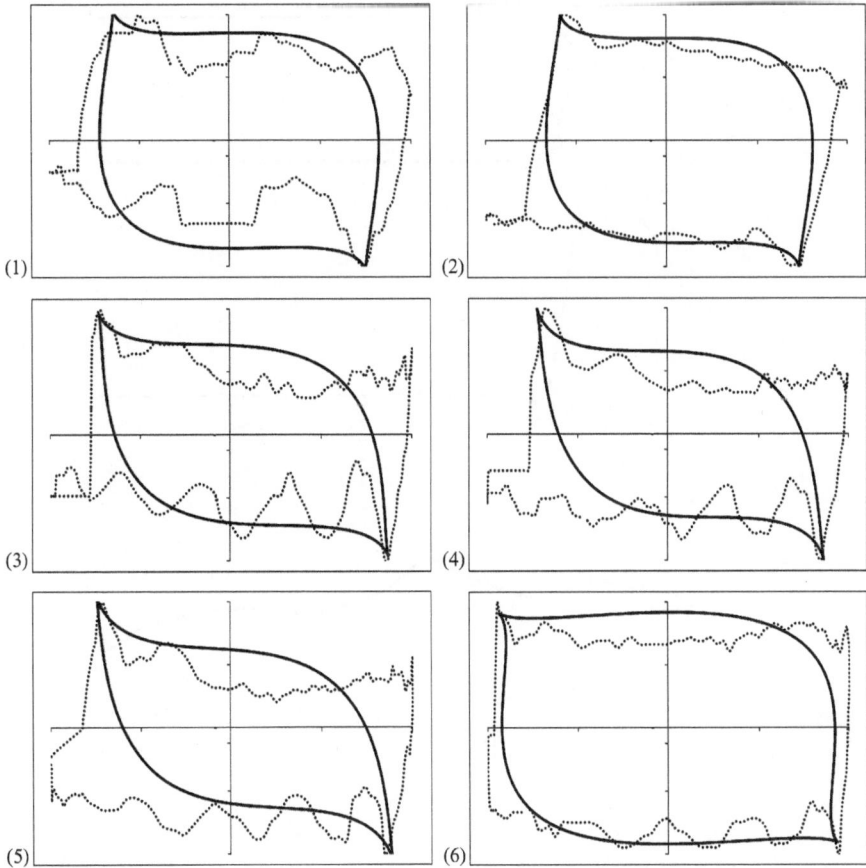

Fig. 6.28 The modeling of a dynamometer card's evolution in time. The difference between each image is 2000 cycles, i.e. about 8.3 hours. The model was train on historical data. Images (1) and (2) are historical data providing the model with the initial data. Using this, the model predicts images (3) to (5) and indicates that at image (5) we have a problem requiring attention. The maintenance measure is performed and we observe, on image (6), the establishment of operations as they should be. See figure 6.29 for more details.

In conclusion, we note that a recurrent neural network can reliably predict a future fault of a rod pump system via predicting the future model parameters of a mathematical formulation of the dynamometer card. In this example, the prediction could be made 16.7 hours in advance of the problem.

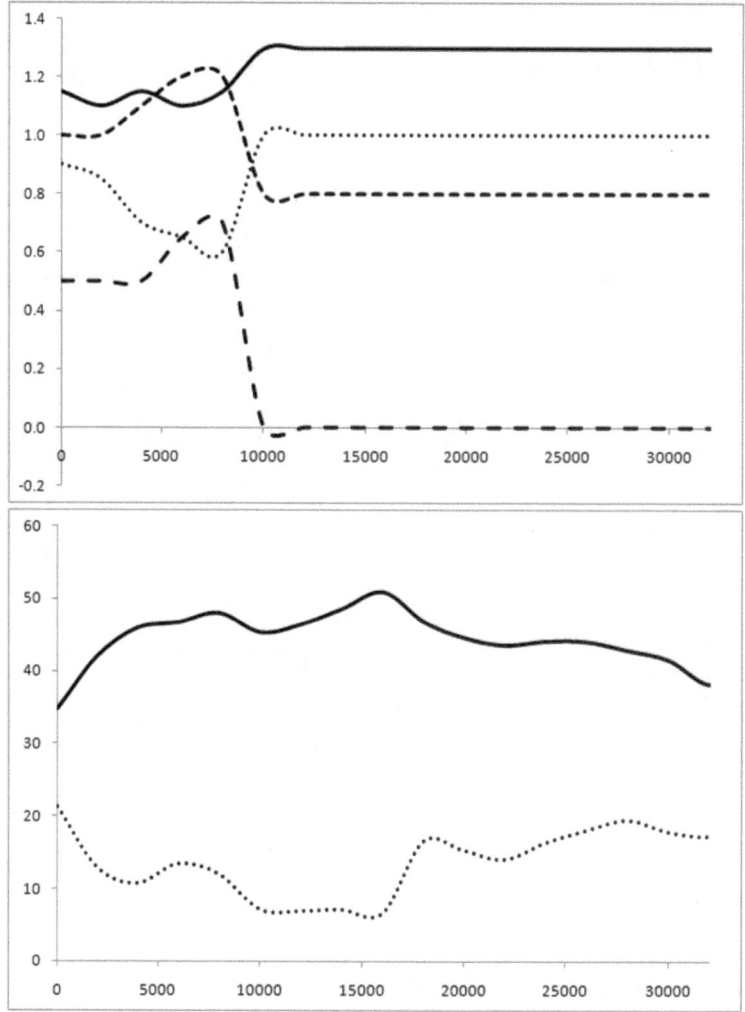

Fig. 6.29 On the top image, we see the evolution of the model's parameters over time: e as the solid line, a as the dotted line, θ as the closely dashed line and c as the long dashed line. All other model parameters were constant throughout. On the bottom we see the evolution of displacement average in the solid line and the displacement width (maximum minus minimum) in the dotted line. The period from time 15000 onwards is to be considered normal operations and so we can observe a gradual worsening of operations leading up to the necessary maintenance measure at time 10000.

Chapter 7
Optimization: Simulated Annealing

There are many optimization approaches. Most are exact algorithms that definitely compute the true global optimum for their input. Unfortunately such methods are usually intended for a very specific problem only. There exists only one general strategy that always finds the global optimum. This method is called *enumeration* and consists of listing all options and choosing the best one. This method is not realistic as in most problems the number of options is so large that they cannot be practically listed.

All other general optimization methods derive from enumeration and attempt to either exclude bad options without examining them or only examine options that have a good potential based on some criteria. Particularly two methods, genetic algorithms and simulated annealing, have been successful in a variety of applications. The later advent of genetic algorithms stole the limelight from simulated annealing for some time [102]. However, from direct comparisons between these two approaches it appears that simulated annealing nearly always wins in all three important categories: implementation time, use of computing resources (memory and time) and solution quality [54, 102]. Thus, we shall focus on simulated annealing. Having said this, the opinions between these two approaches border on religious fervor.

To do some justice to this debate, we will present genetic algorithms in section 7.1 and then describe simulated annealing for the rest of the chapter. It will become clear where the differences lie.

It should also be mentioned here that several methods exist that are usually presented under the heading of optimization methods but that will be ignored here. Such methods are, for example, conjugate gradient methods or Newton's method. The reason we shall ignore them here is that they rely on the problem being purely continuous. They cannot deal with some of the variables being discrete. Industrial problem however almost always involve discrete variables as equipment is turned on and off or many be switched in discrete modes or levels. If you meet with a particular problem that can be written in terms of a purely continuous function, then these methods may not be bad to use. However, general optimization methods may be used profitably in this case as well.

P. Bangert (ed.), *Optimization for Industrial Problems*,
DOI 10.1007/978-3-642-24974-7_7, © Springer-Verlag Berlin Heidelberg 2012

7.1 Genetic Algorithms

Genetic algorithms get their name and basic idea from the evolutionary ideas of biology. There is a population of individuals that mate, beget offspring and eventually die. At any one time, there is population of many individuals but over time these individuals change. These individuals, and thus the whole population, have certain characteristics that are important to us. Particularly, they have a so-called "fitness" derived from the statement "In the struggle for survival, the fittest win out at the expense of their rivals because they succeed in adapting themselves best to their environment" by Charles Darwin. The fitness is thus identical with the objective function of optimization.

The "purpose" of evolution is to breed an individual with the best possible fitness. In nature, changing generations face changing environmental conditions and so it is likely that we will never reach an equilibrium stage at which the truly fittest possible individual can live. In mathematical optimization however, the environment is the problem instance and so does not change. Thus evolution can reach an equilibrium and this is the proposed optimum. Note here that certain concepts are being turned upside down by the method: The ground state of the problem instance (the least likely state) becomes the equilibrium (the most likely state) of the population. The mechanism that achieves this switch is evolution. The fact that the search for something rare is turned into a process of equilibration to something common is the whole point behind genetic algorithms.

The basic features of the genetic approach are thus: The initialization of a population of candidate solutions, the decision of how many such solutions there should be in any one generation, the method for combining several solutions into a new one and the criteria for stopping the search. Note that this approach is a heuristic. We thus have no assurance of finding the true optimum and even if we do find it by chance, we do not have a foolproof way of recognizing it for what it is. This is a pity but we cannot both have our optimum and know it, as it were. This is the price we pay for fast practical execution times.

The decision on how many individuals live per generation is a human design decision that is essentially a black art comparable to choosing the number of hidden neurons in a neural network. The initialization of the first generation is generally done uniformly randomly among all possible candidate solutions. The criteria for stopping have given rise to a significant research field but most applications terminate the search when solution quality no longer improves significantly over many generations, i.e. a convergence criterion. If this seems too haphazard, simply restart the process with a different initial population a few thousand times and take the best outcome. This "restart" method has been shown to provide a small but significant improvement in general and is worth doing for nearly applications with the only exception being if you are very pressed for time (e.g. real-time applications).

The only point that is really complex is defining how solutions mate and beget child solutions. The idea again derives from biological evolution. Two DNA codes seem to combine to make a new DNA code by selecting genes from either parent DNA and then subjecting the result to some mutation.

Supposing that the two parents are the two solution vectors A and B, then we may construct a new solution C

$$A = [a_1, a_2, a_3, \ldots, a_n] \tag{7.1}$$
$$B = [b_1, b_2, b_3, \ldots, b_n] \tag{7.2}$$
$$C = [c_1, c_2, c_3, \ldots, c_n] \tag{7.3}$$

by putting randomly either $c_i = a_i + \varepsilon_i$ or $c_i = b_i + \varepsilon_i$ where ε_i is a small randomly generated mutation term. Choosing an element from either A or B is called the *crossover* operator and adding a small random element is called the *mutation* operator. It is the mutation operator that allows the new solution to be made up of elements that did not exist yet in the initial population and this is crucial in order to gradually visit all possible points.

The method by which we choose elements from A or B can become arbitrary complex or may be a simple 50-50 random choice. Also the method to apply the mutation may be complex or simple. In general, it can be said that only simple problems can be solved by simple methods for these two stages of the solution generation process.

For complex problems, we must prefer certain features in the crossover operation and must gradually suppress mutation over the long term so that the overall solution accuracy can focus on a precise final result. How exactly this is done would go beyond the scope of this book since we wish to focus on simulated annealing but the ideas presented for simulated annealing may be applied to this context.

7.2 Elementary Simulated Annealing

When an alloy is made from various metals, these are all heated beyond their melting points, stirred and then allowed to cool according to a carefully structured timetable. If the system is allowed to cool too quickly or is not sufficiently melted initially, local defects arise in the system and the alloy is unusable because it is too brittle or displays other defects. Simulated annealing is an optimization algorithm based on the same principle. It starts from a random microstate. This is modified by changing the values of the variables. The new microstate is compared with the old one. If it is better, we keep it. If it is worse, we keep it with a certain probability that depends on a 'temperature' parameter. As the temperature is lowered, the probability of accepting a change for the worse decreases and so uphill transitions are accepted increasingly rarely. Eventually the solution is so good that improvements are very rare and accepted changes for the worse are also rare because the temperature is very low and so the method finishes and is said to converge. The exact spelling out of the temperature values and other parameters is called the cooling schedule. Many different cooling schedules have been proposed in the literature but these have effect only on the details and not the overall philosophy of the method.

The initial idea came from physics [95]. The physical process of annealing is one in which a solid is first heated up only to be cooled down slowly in an effort to find its ground state. We are asked to cool the solid so slowly that it remains at thermal equilibrium throughout the entire process. Based on this assumption, statistical physics has been able to calculate the probability distribution of microstates (exact atomic configuration) giving rise to an observed macrostate (the global features of the whole solid). This distribution says that the probability of the solid making a transition between a state of energy E to a state of energy E' (with $E' > E$) is proportional to

$$\exp\left(-\frac{E' - E}{kT}\right)$$

where k is a constant and T is the temperature of the solid. As annealing is an actual physical process used in many manufacturing plants, the computerized simulation of this process is known as *simulated annealing*. In the context of a general substance the method takes the following form [13]:

Data   : A candidate solution S and a cost function $C(\mathbf{x})$.
Result   : A solution S' that minimizes the cost function $C(\mathbf{x})$.
$T \leftarrow$ starting Temperature
while *not frozen* **do**
⎢   **while** *not at equilibrium* **do**
⎢   ⎢ $S' \leftarrow$ perturbation of S.
⎢   ⎢   **if** $C(S') < C(S)$ *or selection criterion* **then** $S \leftarrow S'$
⎢   **end**
⎢ $T \leftarrow$ reduced temperature
end

Algorithm 1: General Simulated Annealing

We immediately see some rather enticing features: (1) Only two solutions must be kept in memory at any one time, (2) we must only be able to compare them against each other, (3) we allow temporary worsening of the solution, (4) the more time we give the algorithm, the better the solution gets and (5) this method is very general and can be applied to virtually any problem as it needs to know nothing about the problem as such. The only points at which the problem enters this method is via creating a perturbed solution and via making a comparison of cost function values.

Note that the inner loop gives rise to a Markov chain as each new state depends only upon its predecessor. Also note that this formulation of the method is quite general. Several pieces are missing: (1) a method for assigning an initial temperature, (2) a definition of "frozen," (3) a definition of "equilibrium," (4) a selection criterion and (5) a method to calculate the next temperature. The cost function and perturbation mechanism are given by the problem we are trying to solve. It must be said that there are, in general, various ways in which perturbations could be generated. The solution quality (final cost and computation time) depends on this choice. As this is highly problem specific, we can only hint at this in section 7.5.

Giving definite computable functions for the five elements named above is referred to as a *cooling schedule* and must be found to turn simulated annealing into an algorithm that can be implemented. As presented above, it may be considered a general computational paradigm but not yet an algorithm.

First we give an example in the form of the traveling salesman problem in which we try to find the optimal journey between 100 cities arranged on a circle beginning from a randomly generated ordering. We chose a large number as initial temperature haphazardly, told the algorithm to stop once no improvement was seen over two consecutive temperatures, defined equilibrium as 1000 proposed transitions and cooled our system by multiplying the current temperature by 0.9. This extremely simple schedule led to the optimal solution of this problem in 35 temperatures. The cost for the problem is the Euclidean distance between the points on the journey and the moves are simple too: (1) reverse a sub-path and (2) replace two sub-paths between towns A and B and C and D by two paths between A and C and B and D [86].

This is an example in which we know what the optimal solution is (a circular journey) and thus we can be happy with this simple cooling schedule. In general however, we do not know what the optimal solution is and so choosing a cooling schedule becomes a problem in its own right because we cannot verify whether the final answer is actually good. In fact, most authors make a rather haphazard choice of functions and parameters for their cooling schedule.

7.3 Theoretical Results

In the limit of infinitely slow cooling, simulated annealing finds the optimal solution for any problem and any instance [123]. This is the central result on which most authors justify their use of simulated annealing. The question of how slowly is slow enough in practice is a complex one that again raises the question of a cooling schedule. It is possible to prove polynomial-time execution of simulated annealing for a large class of cooling schedules [124].

If R is the set of all possible microstates and q_k the stationary probability distribution of the Markov chain (the inner loop of the algorithm), we may define the expected cost $\langle C(T) \rangle$, the expected square cost $\langle C(T)^2 \rangle$, the variance in the cost at equilibrium $\sigma^2(T)$ and the entropy at equilibrium $S(T)$ all at a certain value of the temperature T by

$$\langle C(T) \rangle = \sum_{k \in R} C(k) q_k(T); \tag{7.4}$$

$$\langle C(T)^2 \rangle = \sum_{k \in R} C^2(k) q_k(T); \tag{7.5}$$

$$\sigma^2(T) = \langle C(T)^2 \rangle - \langle C(T) \rangle^2; \tag{7.6}$$

$$S(T) = \sum_{k \in R} q_k(T) \ln q_k(T). \tag{7.7}$$

Furthermore, the specific heat of the system is given by

$$h = \frac{\partial}{\partial T} \langle C(T) \rangle = T \frac{\partial}{\partial T} S(T) = \frac{\sigma^2(T)}{T^2}. \tag{7.8}$$

The expected cost and expected square cost can be shown to be approximated by their averages over the Markov chain by virtue of the central limit theorem and the law of large numbers [3]. These quantities prove helpful not only in analogy to the physics origins of the paradigm but also in providing us with a good cooling schedule. Furthermore, they become important in a discussion of the typical behavior of simulated annealing [124].

In physical annealing, the substance effectively undergoes slow solidification after it has initially been melted at high temperature. It thus undergoes a phase transition. We would expect the total entropy of the system to drastically decrease during this transition and only slowly on either side of it. Physically, such transitions are usually fast but do take a finite amount of time. If we monitor the average energy, standard deviation of the energy and the specific heat of the alloy during the physical annealing process, we should be able to see the phase transition clearly.

Surprisingly, we see the same effects in the evolution of combinatorial problems using simulated annealing. For the traveling salesman problem on 100 cities on the circle, we monitored these quantities over 72 temperatures and plotted them relative to the logarithm of the temperature, see figure 7.1. We see that there is a clear phase transition that is extended over a few temperatures and that both cost and standard deviation vary relatively little on either side of this phase transition. Subject to the assumption that cost is distributed normally over configuration space, we are able to prove that at large temperatures, the standard deviation is constant and the cost inversely proportional to the temperature. At low temperatures, both standard deviation and cost are linearly depended on temperature [2, 70]. This is borne out by the data we have collected.

The specific heat of the instance is roughly constant with a few exceptions. The specific heat of a material body is the amount of heat necessary to be added to the body to increase its temperature by one degree Kelvin per kilogram. It differs in value depending on whether the pressure or the volume of the body are kept constant throughout the process of heating. However, it is a constant property of the material of the body. In the context of combinatorial problems, we could interpret the pressure to be the external forces of change (i.e. the probability distribution of accepting proposed transition) which is constant over the Markov chain that is used to compute the specific heat. The volume of the problem could be considered to be

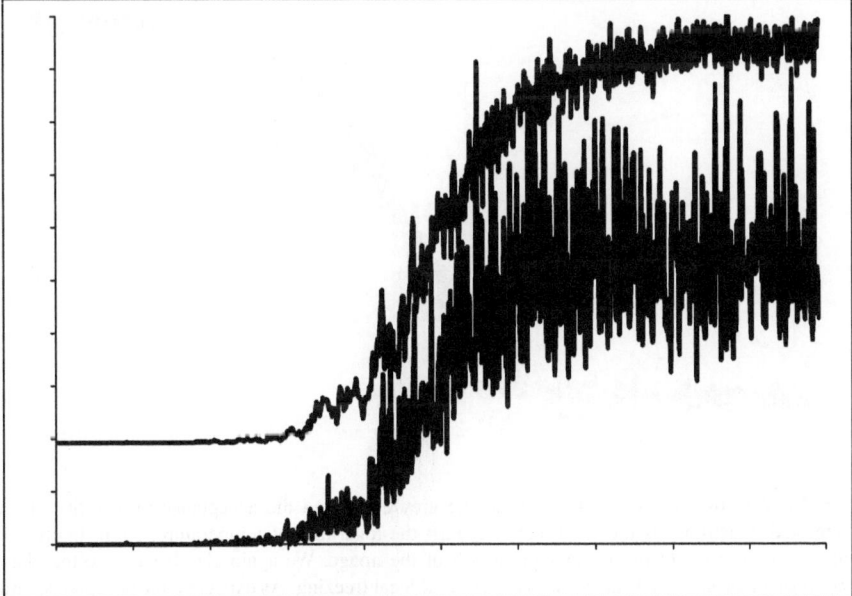

Fig. 7.1 We see the normalized average cost (top curve) and normalized standard deviation (lower curve) against the logarithm of the temperature. As temperature decreases over time, this means that time runs from the right of the plot to the left. We clearly observe the phase transition in the middle of the image.

the average cost during the Markov chain. Thus we are computing the specific heat at constant pressure. In further analogy to physics, we would assume that the specific heat is a constant throughout the execution of simulated annealing. In physics, local maxima in the specific heat indicate local freezing and hence local cluster formation. This is detrimental to finding the ground state of the material and thus should be avoided. In other words, local maxima in specific heat indicate deviation from equilibrium and thus too rapid cooling. If the specific heat at the end of any particular Markov chain is significantly greater than the specific heat computed in the first Markov chain, we should thus disregard the last chain and cool the system more slowly. This adaptive philosophy to a decrement formula will force the specific heat over the evolution to be roughly constant. In our specific example, we see a specific heat maximum around the onset of the phase transition on figure 7.2. Had we cooled more slowly here, we would have in general obtained a much better result.

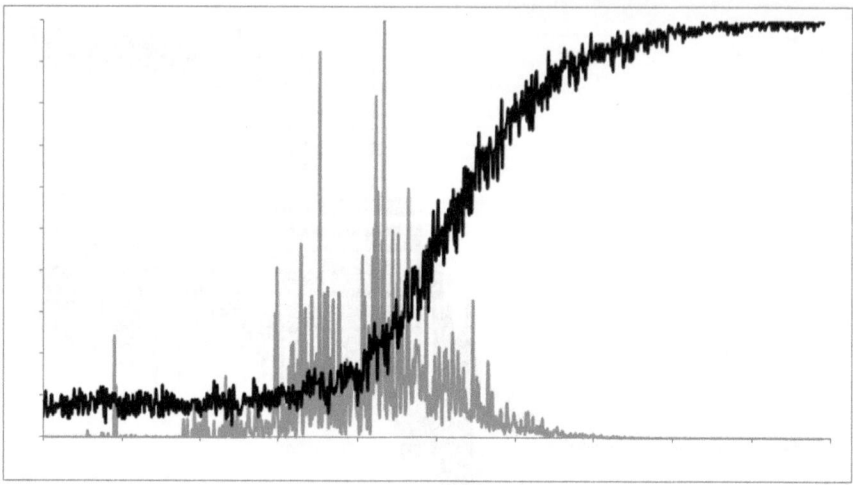

Fig. 7.2 We observe the specific heat as the grey curve and the acceptance probability of the suggested transitions as the black curve against the logarithm of temperature. As in figure 7.1, time therefore runs from the right to the left of the image. We again clearly observe the phase transition in the specific heat curve as the onset of local freezing. As expected, the acceptance ratio of suggestions declines exponentially and, upon becoming too low for further progress, leads to the end of the optimization run.

7.4 Cooling Schedule and Parameters

A host of experimental data from a variety of combinatorial problems show that the performance in both quality of final solution and execution time is highly dependent upon the cooling schedule [124]. We will spend some time reviewing different possibilities for such a schedule and, as we shall see, the parameters that the parts of the cooling schedule require us to choose are no less significant for the performance of the algorithm.

Generally, the quality of the final solution of simulated annealing can compete favorably with the very best of tailored algorithms for specific combinatorial problems however at the cost of additional execution time [124]. This observation has been made in many papers but all of them have used very simple cooling schedules and have not tuned the parameters of these schedules well. This leaves us to speculate that one may expect additional quality and time performance from simulated annealing after appropriate tuning. If this were consistently true, this method may beat a number of tailored algorithms in quality.

There are parallel implementations of simulated annealing that speed up the execution considerably. These are, however, more complex and deviate somewhat from the original physical and intuitive ideas. They are also harder to implement and so we refer to the literature for this, e.g. [100] and references. We begin the discussion of cooling schedules with the cautionary remark that an experimental comparison of the major cooling schedules (with tuned parameters) has never been done and so advantages and disadvantages are a matter of theory for now.

7.4.1 Initial Temperature

The starting temperature has to be chosen such that the system can make highly uneconomical transitions in the beginning and later settle down to temperatures at which very few such transitions are possible. Thus, in combination with the temperature decrement formula and the selection criterion, this temperature should be chosen high enough but not too high. There are two major directions in which investigators have chosen to go.

The first is to say that the initial temperature T_0 should be such that a certain percentage χ_0 of uneconomical transitions are accepted [112, 79]. We start by assuming a Gaussian distribution of cost fluctuations because this is our generic selection criterion and thus set the probability of acceptance that we want (χ_0) equal to the probability at the initial temperature [57],

$$
\chi_0 = \exp\left(-\frac{\overline{\Delta C^{(+)}}}{T_0}\right) \to T_0 = \frac{\overline{\Delta C^{(+)}}}{\ln\left(\chi_0^{-1}\right)}
$$

where $\overline{\Delta C^{(+)}}$ is the average of all cost function increases observed. In order to use this formula, we thus perform one Markov chain in order to compute $\overline{\Delta C^{(+)}}$ and then compute T_0 by choosing some χ_0. Estimates of this kind are used in many papers [84, 85, 119, 66, 97]. While this is a relatively simple and utilitarian solution of finding a good starting temperature, most application oriented papers do not even go this far but merely decide to fix T_0 to a number that seems to give good results empirically. Once this number is found using a few test instances, this number is fixed for all subsequent instances and thus becomes part of the problem specification. Clearly this is not a good choice of strategy. In the best case, there will be many instances for which the computation time taken will be larger than needed but it is likely that in many cases the final solution found will be inferior to the one that could have been found with a different initial temperature. We thus advise on an adaptive tuning of the initial temperature according to some model.

A more sophisticated approach is based on the assumption that the number of solutions corresponding to a particular cost C, the *configuration density*, is normally distributed with a mean of $\overline{C}$ and standard deviation of σ_∞. These parameters are empirically determined during a Markov chain. We may then compute the expectation value of the cost as a function of the temperature $\langle C(T)\rangle \approx \overline{C} - \sigma_\infty^2/T$, which is a local Taylor expansion where $\overline{C}$ is an average taken at the current temperature, i.e. over the current Markov chain. The variance $\sigma_\infty^2 = \langle C(T)^2\rangle - \langle C(T)\rangle^2$ is estimated to be $\sigma_\infty^2 \approx \overline{C(T)^2} - \overline{C(T)}^2$. Finally, we agree that we would like the initial expectation of the cost to be within x standard deviations from the average cost. Together with the empirical estimators for the expectation values, we obtain [128]

$$T_0 > x\sqrt{\overline{C^2} - \overline{C}^2}.$$

We learn in statistics that 68% of all cases lie within one, 95% within two and 99.7% within three standard deviations of the mean for a normal distribution. The number of standard deviations x, thus again comes down to choosing a χ_0. From the cumulative Gaussian probability distribution the number of standard deviations and the initial acceptance probability are thus related by

$$\chi_0 = \int\limits_{-x\sigma_\infty}^{x\sigma_\infty} \frac{1}{\sqrt{2\pi}\sigma_\infty} \exp\left(-\frac{\left(C-\overline{C}\right)^2}{2\sigma_\infty^2}\right) dC.$$

This is analogous to a practical method at which we start the metal off at room temperature and heat it up gradually until we believe it is hot enough for it to be well mixed (some distance past its melting point for instance) and then we begin the annealing process in earnest. It is the concept of *melting point* that we have essentially attempted to describe in this section and that accurately captures what we need the initial temperature to be.

7.4.2 Stopping Criterion (Definition of Freezing)

The analogy between the melting point of a metal to be annealed and the starting temperature of a combinatorial problem to be simulated annealed was drawn in the previous section. This can be continued between the freezing point of a metal and the stop criterion of the simulated annealing process. Physically speaking, the freezing temperature and the melting point are the same (this temperature marks the phase transition between the solid and liquid phases) but this transition is not instantaneous. In the context of optimization, the starting temperature is higher and the final temperature lower. Thus, a real phase transition does not occur at one instant but over a period of time. The entropy does not have an actual discontinuity (as the theory would have us believe) but rather it has a sudden and drastic change over a small but finite time frame.

The simplest proposal for the final temperature is to fix the number of the different temperatures and thus the final temperature depends upon the temperature reduction formula. The actual number varies between six [113] and fifty [47] in the literature. The analogue of waiting until no more consequential transitions are made is to wait until the optimal configurations found after a number of Markov chains (typically the last four) are identical [79, 97, 115]. We may further require that the probability of accepting a random transition is smaller than some fixed value χ_f by analogy to the treatment of the starting temperature [57].

A number of more sophisticated proposals are made in the literature. Suppose that we are at a local minimum C_0 in the cost function and the lowest cost value of any configuration in the neighborhood of the current configuration (i.e. reachable by a single transition) is C_1. Then we would like that the probability of transiting from the local minimum to this point should be lower than $1/R$ where R is the size of the neighborhood. This condition (assuming that cost is normally distributed, as before) gives [128],

$$T_f \leq \frac{C_1 C_0}{\ln R}.$$

The choice that $1/R$ is to be the cut-off probability seems reasonable but nevertheless we may consider this a general statement under the Gaussian assumption and input any desired probability.

Alternatively, we may require that the probability of the last configuration reached in a Markov chain being more than ε (in cost) over the true minimum of the cost function, is less than some real number θ. This may be used to derive the condition [89]

$$T_f \leq \frac{\varepsilon}{\ln(|\mathscr{R}| - 1) - \ln \theta}$$

where $\mathscr{R}$ is the set of all possible configurations. This obviously suffers from having to choose an ε and a θ (similar to the χ_f above) and having to calculate the size of the configuration space.

Consider the difference between the average cost $\overline{C}(T_k)$ during the kth Markov chain and the true optimum. This may be expanded as a first-order Taylor series when T_k is small. This difference (calculated by the Taylor series) relative to the average cost in the first Markov chain is desired to be lower than some fixed ε divided by the terminal temperature T_f. Finally this gives [4, 101]

$$T_f > \frac{\overline{C^2}(T_f) - \overline{C}^2(T_f)}{\varepsilon \left(\overline{C}(T_0) - \overline{C}(T_f) \right)}$$

7.4.3 Markov Chain Length (Definition of Equilibrium)

Continuing our physical analogy, we are to anneal our substance at equilibrium. Thus we are only allowed to lower the temperature when the substance has reached thermal equilibrium at the current temperature. We need a firm definition of this concept in combinatorial terms in order to terminate our Markov chain. A strict mathematical definition of equilibrium is practically impossible as it would entail computing the probability distribution of configuration space which would correspond to the simplest of all optimization algorithms (check all possibilities and choose the best one).

Let the length of the kth chain be L_k. The simplest practical way is to give a definite fixed length to every chain so that L_k is independent of k and depends only upon the problem size, i.e. it is some polynomial-time computable function of $|\mathscr{R}|$, the size of the configuration space. Many authors simply choose some number, e.g. $L_k = 100$ [47, 46, 90, 51, 115]. Alternatively we use this function only as a ceiling for the length and terminate the chain possibly before, such that the number of accepted transitions is at least some η_{min} (this requires a ceiling because at low temperatures the acceptance ratio is lower) [79, 57, 84, 85, 97]. On the other hand, we may want to require that the refused transitions are at least a certain number [113]. This seems counter-intuitive as this leads to shorter chains at low temperatures and thus a speedup of cooling whereas one would assume that achieving equilibrium takes longer at low temperatures.

More physically, consider breaking the chain into fixed-length (in terms of accepted transitions) segments. The cost of a segment is the cost of the last configuration. When the cost of current segment is within a cut-off of the preceding segment, we terminate the chain [119, 58, 52]. This is more intuitive because this is related to the fluctuations in cost over the chain. We terminate when the fluctuations settle down; a definition of equilibrium that an experimental physicist might agree with.

Statistically speaking, we would wish for a sufficiently large probability to make an uneconomical transition (possibly out of a local minimum) to be maintained throughout the chain. A reasonable estimate, based on Markov chains, for a scale length of a specific chain is

$$N \approx \frac{1}{\exp\left(-\left(C_{max}C_{min}\right)/T\right)}$$

where C_{max} and C_{min} are the largest and smallest cost observed so far (including previous chains). This is further corroborated by the fact that N plays a similar role in stochastic dynamical systems as the time constant plays in linear dynamical systems; it is thus really a length scale [59, 108]. Taking the actual length of the chain to be a few Ns should thus be enough to get to equilibrium; exactly how many, however, remains to be decided by the user.

Finally, it is possible to show that the number of accepted transitions within an interval $\pm\delta$ about the average cost $\overline{C}$ reaches a stationary value

$$\kappa = \mathrm{erf}\left(\frac{\delta}{\sigma(T)}\right) \approx \mathrm{erf}\left(\frac{\delta}{\sqrt{\overline{C^2}-\overline{C}^2}}\right),$$

where $\mathrm{erf}(\cdots)$ denotes the error function, at equilibrium and we may take this as our definition (practical care has to be taken to avoid extremely long chains at low temperatures) [94].

7.4.4 Decrement Formula for Temperature (Cooling Speed)

After each Markov chain the temperature is decreased in analogy to the physical annealing process in which a metal is cooled at equilibrium. The new temperature T_{k+1} is calculated from the old temperature T_k very simply by either keeping their ratio [79, 57, 47, 46, 90, 51, 84, 85, 97, 115] or their difference [119, 66] a global constant. The ratio is usually taken between 0.9 and 0.99 but also 0.5 has been used. If the difference is used, then it is determined by fixing the number of different temperatures, the initial temperature and the final temperature. The ratio rule is used very frequently in applications to the virtual exclusion of other rules.

It is the decrement rule that has the most impact upon the quality and efficiency of the algorithm among the five rules that we must prescribe [124]. We would like to decrease the temperature slowly so that the subsequent chains do not have to be too long in order to reestablish equilibrium but not too slowly so that the algorithm takes too much time to freeze.

We begin with the reasonable statement that the stationary probability distributions of two successive temperatures should be close, i.e. their ratio larger than $1/(1+\delta)$ and smaller than $1+\delta$ for some (small) real number δ. Depending on subsequent assumptions, we may derive the following rules,

$$T_{k+1} = T_k \left(1 + \frac{\ln(1+\delta)}{3} \frac{T_k}{\sigma(T_k)} \right)^{-1} , \text{ see [4, 1]} \tag{7.9}$$

$$T_{k+1} = T_k \left(1 + \frac{\ln(1+\delta)}{3} \left(\frac{T_k}{\sigma(T_k)} \right)^{3} \right) , \text{ see [82]} \tag{7.10}$$

$$T_{k+1} = T_k \left(1 + \frac{\gamma T_k}{U} \right)^{-1} , \text{ see [89]} \tag{7.11}$$

$$T_{k+1} = T_k - \frac{T_k^2 \ln(1+\delta)}{C_{max} + T_k \ln(1+\delta)} , \text{ see [101]} \tag{7.12}$$

where γ is some small real number and U is an upper bound on the difference in cost between the current point and the optimum.

Alternatively, we can begin from the statistical mechanics formula for specific heat and approximate, as we have done so far, the expected cost by the average cost. This leads to

$$T_{k+1} = T_k \exp \left(-\frac{\lambda T_k}{\sigma(T_k)} \right)$$

where λ is the number of standard deviations by which the average costs of different chains are allowed to differ; we require $\lambda \leq 1$ [94].

7.4.5 Selection Criterion

Generally, the Maxwell-Boltzmann distribution is assumed to be a reasonable criterion for accepting or rejecting a proposed transition for the worse by the analogy to statistical physics and so the probabilistic selection criterion is relative to the function $\exp[-\Delta C/T]$. The discussion of whether a different condition should be chosen is based on the observation that transitions of high cost difference can help to get the system out of local minima and these are accepted rather less often at low temperatures. Furthermore, at large values of the temperature virtually all transitions are accepted without bias. One may wish to bias the selection of transitions such that large transitions are more likely at lower temperatures to help approach the optimum faster. There are many possible choices but they are all centered around attempting to force faster convergence and not lower final cost. It is our experience that with present hardware, it is not necessary to speed up the algorithm (at the cost of possibly a worse solution) for almost all practical problems.

We mention one simple way to tune the selection process, namely the reintroduction of Boltzmann's constant in the Maxwell-Boltzmann distribution, i.e. changing the function to $\exp[-\Delta C/kT]$ where k is a constant. In physics, this constant takes one universal and constant value; it can thus be thought of as a conversion factor of Kelvins to Joules (units of temperature to units of energy). In combinatorics, this factor would convert units of temperature (whatever that may mean here) to units of cost. In our discussion of initial temperatures and decrement formulas, however the unit of temperature was the same as the unit of cost and so the constant, in this context, does not need to convert units. Specifically, there is some evidence that $k = 2$ may lead to slightly faster convergence to equilibrium [68].

A very interesting approach uses the so-called Tsallis statistics to attempt a speed up of annealing's convergence without a loss of quality. This is very promising but is beyond the scope of this book to discuss, see [80] for a start.

7.4.6 Parameter Choice

We have seen that we must choose five elements (initial and final temperatures, chain length, decrement formula and selection criterion) to turn the paradigm of simulated annealing into an algorithm in addition to formulating our problem in terms of transitions. This choice is far from obvious. In addition, almost all of these elements depend on some parameters that we must also choose. We have some theoretical and practical guidelines and intuition as to what rules to choose but the parameters generally escape precise quantification by intuition. We are thus lead to the question: Do slight variations in the parameters make measurable differences in the performance (quality and speed) of the simulated annealing algorithm applied to a particular problem? The experimental answer is definitely affirmative. Thus we have to make intelligent choices.

It is unfortunate that almost all practitioners of the simulated annealing paradigm do not put much effort into finding the optimal parameters. From the literature it seems that the vast majority follows the following method. A few test instances of the problem are generated. Some of these have known optimal solutions and some do not. The parameters are adjusted *manually* such that the known cases are solved to optimality and the unknown cases are solved to a final cost that seems reasonable in the light of the known cases. This manual adjustment means in practise that rather few different parameter sets are tried and the first one that looks good in the above way is kept. Furthermore, the parameters are then kept fixed for all future cases to be solved and are hardly ever (except perhaps in the case of the initial temperature) varied. However an optimal parameter set can improve the average solution quality appreciably over a manually chosen one.

Another method used more recently is to try a few values for each parameter and then use linear regression to obtain some optimal interpolated parameter set [63]. This is essentially a manual adjustment as well as there is no good method to choose the few sets on which the regression is based.

Alternatively, we can regard simulated annealing as a function of its parameters that returns the relative cost reduction $\alpha = \left(C_{final} - C_{initial} \right) / C_{initial}$ of the problem instance. This is not quite good enough because α has a probability distribution that is largely unknown. However, if we generate a large number of similar instances of the same size and compute the average cost reduction $\overline{\alpha}$ by running simulated annealing with identical parameters for each one, then it should return the expectation value of the relative cost reduction. This is a good measurement of the efficacy of the method and we shall take this as our figure of merit function to find the optimal parameters. In short, we have a multidimensional function minimization problem (maximize $\overline{\alpha}$ is the same as minimizing $1/\overline{\alpha}$).

The simplest version of simulated annealing sets five constants A, B, C, D and E to some initial values and looks like this:

Data : A candidate solution S and a cost function $C(\mathbf{x})$.
Result : A solution S' that minimizes the cost function $C(\mathbf{x})$.
$T \leftarrow A$
while $T > B$ **do**
    for $i = 1$ *to* C **do**
        $S' \leftarrow$ perturbation of S.
        if $C(S') < C(S)$ *or Random* $< \exp[(C(S') - C(S))/DT]$ **then** $S \leftarrow S'$
    end
    $T \leftarrow ET$
end

Algorithm 2: Simple Simulated Annealing

In words, we start with a constant temperature A and define a constant temperature B to be the freezing point. Equilibration is assumed to occur after or within C steps of the proposal-acceptance loop where the selection criterion is the thermodynamic Maxwell-Boltzmann distribution with Boltzmann's constant D after which the temperature is decremented by a constant factor E. The standard choices for these constants are $A = C(S)$, B is 100 times smaller than the best lower bound on cost, $C = 1000$, $D = 1$ and $E = 0.9$. After successful implementation of this algorithm, one usually plays with these parameters until the program behaves satisfactorily. It is clear that implementing this method is very fast and we observe from the literature that the vast majority of applications are computed using the version of simulated annealing given in algorithm 2 where the five parameters are determined manually [54].

Using this interpretation, we may regard simulated annealing as defining a function $\overline{\alpha} = \overline{\alpha}(A,B,C,D,E)$ depending on five parameters (for the simple schedule). We would like this average reduction ratio to be as large as possible.

This is yet another optimization problem with a function instead of a combinatorial problem. We are able to evaluate the function only at considerable computational cost (N runs of simulated annealing for N randomly generated initial configurations) and we do not know its derivative accurately. Even approximating the derivative comes only at heavy computational cost. There are many optimization methods such as Newton's method or more generally a family of methods known under the names Quasi-Newton or also Newton-Kantorovich methods that rely on computing the derivative of the objective function. Some of them require high computational complexity due to the computation of the Hessian matrix but complexity considerations are secondary here. The most important reason against all these methods is that the derivative computation is not very accurate for the function constructed here and this loss of accuracy in an iterative method would yield meaningless answers. Indeed, such methods were tried and the results found to be unpredictable because of error accumulation and much worse than the results obtained by methods not requiring the computation of derivatives. The method of choice for optimizing a function over several dimensions without computing its derivative is the downhill simplex method (alternatively one may use direction set methods). Thus, we use the downhill simplex method to minimize $\overline{\alpha}(A,B,C,D,E)$.

The starting point for the simplex method will be given by those values of the five parameters that we obtain after some manual experiments. This is done for the reason that most practitioners of the simulated annealing paradigm choose their parameters based on manual experiments [54]. The other points on the simplex are set by manually estimating the length scale for each parameter [127].

We find, after extensive computational trials on a variety of test problems, that the average improvement in the reduction ratio after annealing has been parametrized by the downhill simplex method as opposed to human tuning is 17.6%. We believe this is significant enough to seriously recommend it in practise. Note that this is an average and so there are cases where the improvement is small and cases where it is large. It seems impossible to tell *a priori* what the result will be.

Many simulated annealing papers have been published that center around the topic of performance of the algorithm in terms of getting to an acceptable minimum *quickly* [102]. A variety of cooling schedules have been designed that can reduce the computation time at the expense of solution quality. While the author experimented with a number of open-source implementations of simulated annealing for a variety of optimization problems with tools such as a profiler, speed-ups of up to three orders of magnitude were achieved. This is in contrast to claimed speed-up factors of between 1.2 and 2.0 that come from changing the cooling schedule *at the expense of solution quality* [102]. Thus the author believes the speed of the simulated annealing method to be so dominated by programming care that he has not attempted to simultaneously optimize solution quality *and* execution speed. This simultaneous optimization would, however, be no problem in principle after one made the, completely random, decision how relatively important speed is in relation to quality.

Thus we may draw a number of conclusions that would appear to hold in general: (1) The solution quality obtained using simulated annealing depends strongly on the numerical values of the parameters of the cooling schedule, (2) the downhill simplex method is effective in locating the optimal parameter values for a specific input size, (3) the parameters depend strongly on input size and should therefore not be global constants for all instances of an optimization problem, (4) the improvement in solution quality can be significant for theoretical and practical problems (up to 26.1% improvement was measured in these experiments which is large enough to have significant industrial impact).

Furthermore, the reason that the usual manual search is so much worse than an automated search seems to be that the solution quality (as measured by the average reduction ratio) depends strongly on the cooling schedule parameters, i.e. the landscape is a complex mountain range with narrow valleys that are hard to find manually. Finally, the improved schedule parameters, in general, lead to slightly greater execution time but in view of the dramatic improvement of quality (as well as the fact that execution time seems to be dominated by programming care) this is well worth it. However the computation times are generally so low nowadays with powerful computers that increasing the speed of annealing at the expense of quality is a non-issue. Rather we would expand the computation time for the benefit of additional quality.

7.5 Perturbations for Continuous and Combinatorial Problems

Apart from the cost function, with which we compare any two proposed solutions, the only other point in annealing that the problem enters the algorithm is in the method to perturb or change any proposed solution.

This method to modify a solution must be carefully constructed such that we have a good chance to meet with many solutions and to be able to exit local minima. Let us imagine that we are dealing with a continuous problem. That is, a problem in which the independent variables (the one whose value we want to determine) take on continuous values as opposed to discrete values. Then a solution is any value of the independent variable vector $\mathbf{x}$ that obeys the boundary conditions. In order to generate a new vector $\mathbf{x}'$ from this, we can create several ideas

1. Set $\mathbf{x}'$ to a random vector independent of $\mathbf{x}$. This is a simple and intuitive idea but it violates the basic philosophy of simulated annealing of adaptive change.

This method makes annealing essentially a variant of random search (take the best solution of many randomly generated ones) and performs poorly.

2. Set $\mathbf{x}'$ equal to $\mathbf{x}$ and change one element by $\pm\Delta$ for some *a priori* chosen length scale Δ. This is better but performs poorly as well, as most problems tend to have shorter length scales at lower temperatures due essentially to the phase transition observed at intermediate temperatures.

3. Do the above but make Δ into $\Delta(T)$ a function of temperature. The scale should decrease with temperature, that much is clear. There is wide disagreement in the literature as to how to decrease it exactly. Describing these methods would carry us too far afield. We merely note that the length scale of any particular variable at any particular temperature may be estimated by performing many transitions for various Δ for this variable at that temperature and noting down the variation in the cost function thus achieved. In this manner, we may empirically construct a $\Delta(T)$.

4. Do the above but assign a different $\Delta(T)$ to each element in $\mathbf{x}$ because every variable may (and in general will) have its own length scale.

5. Do the above but do not change only one element of $\mathbf{x}$ but several during a single move. We then ask how many is "several?" We have found that about 10% of the number of elements in $\mathbf{x}$ is a good number to vary simultaneously. Which they are should be chosen randomly.

In absence of domain knowledge that may allow us to design a better transition mechanism, the last of the above suggestions has performed best for many experiments of the author. In case of doubt, this approach is recommended.

If we are dealing with a problem that is not continuous but rather discrete, matters are more diverse. We now need to take into account the actual structure of the problem.

Take, for example, the traveling salesman problem. The mechanism that has proven to be the best has two moves: reverse any partial path and interchange two partial paths, e.g. $A \rightarrow B$ and $C \rightarrow D$ with the two partial paths $A \rightarrow C$ and $B \rightarrow D$. There have been many other suggestions for generating a new solution from an old solution but it is this suggestion that has been found to perform best. It is unclear, before experimenting, which set of moves will perform best.

It is apparent from the structure of the moves for the traveling salesman problem, that these were designed with the structure of the problem itself in mind – the problem is about a path between points without repetition. We cannot use these moves for a different problem. As such we must really design a move set with respect to a problem.

There is no general theory for constructing a transition mechanism. We must think what is natural for the problem at hand and try it out. In general several suggestions will have to be tested. Most often we will have no theoretical explanation for the better performance of the winner but must merely be content to observe which happens to be the best.

We note in closing that the suitability of the transition mechanism is a major point in using simulated annealing. If you use a poor transition mechanism, annealing will take much longer (require more transitions) and may indeed converge to a poorer outcome. Note also that the cooling schedule depends on the transition mechanism and so the cooling schedule must be tuned to a particular transition mechanism.

7.6 Case Study: Human Brains use Simulated Annealing to Think

Co-Author: Prof. Dr. Adele Diederich, Jacobs University Bremen gGmbH

Humanity has long searched for the mechanism that allows the human brain to be as successful as it is observed to be in solving a variety of problems both new and old every day. Much is known about the architecture of the brain on the level of neurons and synapses but very little about the global *modus operanti*. We find evidence here that simulated annealing is that elusive mechanism that could be called the 'formula of the brain.' By examining the traveling salesman and capacitated vehicle routing problems that are typical of everyday problems that humans solve, we illustrate that none of the optimization methods known to date match the observations except for simulated annealing. The method is both simple and general while being highly successful and robust. It solves problems very close to optimality and shows fault-tolerance and graceful degradation in the presence of errors in both input data and objective function calculations.

The human brain is constituted of approximately 10^{11} individually simple computational elements (neurons) that are interconnected via approximately $5 \cdot 10^{14}$ synapses[1] [55, 111]. These large numbers prohibit a comprehensive direct computer model of the brain. Even if it were possible, such a model would be essentially epistemological, i.e. it would treat the brain as a "black box" and would concern itself only with input and output to this box. It is eminently more desirable to search for an ontology of the human brain, a theory that (at least to some degree) explains as well as reproduces input–output pairings.

The importance for science in general of understanding how our brain thinks in global terms can hardly be overemphasized. Given the philosophical nature of the issues, it seems unreasonable to be able to resolve the nature-nurture, consciousness-complexity or intelligence debates on such grounds. However, many issues of scientific interest can be tackled from this basis such as the performance issues at the basis of intelligence and all manner of questions regarding memory and learning as well as modularization or compartmentalization of the brain. Furthermore, through better understanding of the human 'hardware' it should be possible to facilitate improved learning, recall and equilibrated and enhanced brain usage. In brief colloquial terms, an ontological brain mechanics forms the essential introduction to a brain operations manual for the scientist as well as for the lay-thinker.

---

[1] Graph-theoretically speaking the brain is a very sparse graph – with 10^{11} nodes, a graph with all possible edges would have $0.5 \cdot 10^{22}$ edges meaning that the human brain has approximately 0.00001% of all possible synapses. This is, of course, necessary as compartmentalization and modularization are quite essential for the myriad functions that the brain has to perform simultenously.

The average human being must make complex choices many times per day. Many of these fall into the category of optimization problems: Choosing the 'best' alternative from a wealth of possibilities; a simple example being that of planning a route between many stops. The meaning of 'best' differs widely between problems[2] but it is clear that we must be able to (and are able to) compare several possibilities as to their goodness when trying to find the best alternative. The process of 'thinking about' the problem (i.e. considering the relative goodness of various possible alternatives) takes time and often we consider an alternative that is worse than the best one found so far in an effort to find an aspect of that alternative that will allow us to find an even better alternative later on – we accept temporary losses in expectation of greater gain at a later time.

In most problems, the total number of possible alternatives is astronomically large and no simply recipe for solution exists. As an example, planning the shortest route between n stops while running errands would have us consider $(n-1)!$ possible routes. Most of these problems can only be (realistically) solved using heuristic methods. The human brain is thus capable of selecting a good alternative from a large set of possibilities *without* considering all possibilities. Additionally, the brain does not search randomly but 'intelligently' considers the alternatives. It is unreasonable to believe that the human brain has separate solution strategies for each possible different problem. This would require a vast brain (which we do not have) and enormous learning (which we do not have time for). Thus there must exist a central problem solving apparatus that manages to solve many very different problems to a reasonable degree each[3]. Furthermore, the neuron-synapse structure of the brain operates approximately 1000 times slower than current computer hardware and humans still regularly outthink computer programs in tasks such as pattern recognition. It is sometimes thought that massive parallelism is the key to this performance gap [117, 111]. *Our thinking strategy is thus problem-independent and very quickly obtains a nearly optimal solution via a directed search through a very small portion of the space of possible solutions.*

We wish to draw a parallel between the SA algorithm and the human brain functionality in solving optimization problems. As most problems encountered in everyday life are optimization problems, we will take this to be a strong indication that the human brain uses SA as its general problem solving strategy. Learning is adopted into this in two ways: (1) The cooling schedule of the SA paradigm is very flexible and amenable to substantial tuning and (2) after sufficient experience with a particular kind of problem, the brain may develop a custom method for dealing with those particular issues important to that human being.

SA is very powerful as we have seen but it is also very robust. Robustness refers to the preservation of the method's ability to find a good solution in the presence of

---

[2] Examples include minimizing the number of kilometers needed for a journey, the amount of time required for a job, the number of trucks needed to supply a chain of stores, the number of rooms required for a conference, the best assignment of employees to job tasks and so on.

[3] Note that due to the astronomically large number of possible solutions and the non-existence of any quick guaranteed solution schemes, the human brain cannot be expected to (and does not) solve these problem to optimality but only close enough for practical purposes.

noise (errors in the input data) and/or uncertainty (errors in the objective function evaluation). Clearly the human brain is very robust and this feature must carry over into any ontological mechanics of the brain.

The proper unit of time for the SA paradigm is the number of proposals made. It is useless to measure the actual time taken as this depends too strongly on the computer and programmer. As each proposal necessitates the construction of a candidate solution, its comparison to the current reference solution and its subsequent acceptance or rejection, we postulate based on timing measurements of the brain that the average human is capable of doing this in 3 microseconds [56, 111]. This means that the human brain considers approximately 333 proposals per second. We shall take this as the conversion factor in order to compare our SA algorithm to brain measurements[4].

In our experiments we will task a number of human subjects and the SA algorithm with a number of instances of the traveling salesman (TSP) and the capacitated vehicle routing problem (CVRP). In both problems, the locations (on the two-dimensional Euclidean plane) of n cities is chosen. In the TSP, the shortest possible round-trip journey meeting each city exactly once is to be constructed. In the CVRP, each city has an associated demand value that the traveling salesman has to meet given that his vehicle has a maximum capacity. A particular city is flagged as the 'depot' where the salesman has to periodically return to refill his vehicle. The CVRP asks for the shortest journey meeting each city (except the depot) exactly once starting and ending at the depot and fulfilling all demands while never exceeding the vehicle capacity[5].

---

[4] It should be noted that almost no proposals considered are done so consciously. The computation power of the brain is vast but relies on almost all of the computing to be done unconsciously.

[5] Method Note: Both problems can, of course, be asked with the cities not on the two-dimensional Euclidean plane but this restriction made it possible to get human test subjects to solve the same instances as the computer. The instances used had between 10 and 70 cities; half the instances were taken from actual cities on the Earth projected onto the plane and the other half were uniformly randomly distributed in a square.

For the human test subjects, the computer screen first informed them how many cities the next problem had and what the maximum time limit was, then a fixation cross was displayed on the screen's center and then the cities as dots on the screen were displayed together with a clock counting backwards from the maximum time limit. The subjects had to use the mouse pointer to click on the cities in the order that they wished to visit them. A choice could be undone by clicking the other mouse button and the click times were all recorded as well as the length of the final journey. It was found that the average subject needs approximately one second per city just to perform the clicking operations, i.e. giving less than this much time would not yield a complete tour of the cities. Each instance was displayed several times for different maximum time values in order to measure the time progress of the subjects as they were given more or less time to think. Each time that the instance was redisplayed it was rotated by a different angle so that the instance did not look the same as it did on its previous display. As such the learning effects of the experiment were kept to the general task and not specific to an instance.

For the computer algorithm, we used a cooling schedule that assesses the correct starting temperature by heating up the system until 99% of all proposals are accepted (using the Maxwell-Boltzmann distribution for accepting cost-increasing proposals). Equilibrium is defined as 200 proposals and the temperature is decreased by a constant factor of 0.99 until the cost does not change over four consecutive equilibria. This schedule is capable of solving to optimality all small

In order to compare the data from the subjects to that of the computer, we note that for the SA paradigm the graph of percentage cost deviation from optimum versus time is largely independent of the problem size for small instances. Furthermore, this graph as well as the graph of cost variance versus time display a smoothed step function profile. The initial plateau is a feature that SA could be criticized for as an optimization algorithm as it could be viewed as a waste of computational resources; after all we only begin to see progress after a substantial amount (roughly one-third) of time has been spent. From a catalog of possible smoothed step function forms [103], we have determined that the best fit is the complementary error function; $a \cdot \mathrm{erfc}(bx+c)+d$ where a, b, c and d are constants and $\mathrm{erfc}(x) = \frac{2}{\sqrt{\pi} \int_x^\infty \exp(-t^2)dt}$. Other optimization algorithms do not have this feature profile; their cost versus time graphs generally follow a decaying exponential graph.

We do indeed find a scale invariance in the human subjects' performance as it was expected from SA experience, i.e. the normalized cost and standard deviation curves do not depend upon the problem size. Thus we restrict ourselves to presenting data from a particular instance. See figure 7.3 where the SA output has been scaled in time according to the rate of 333 proposals per second.

The notable features of this comparison are thus: (1) Scale invariance was observed in both human subjects and computer algorithm, (2) the cost and standard deviation functions agree closely between computer and human subjects, (3) the (independently arrived at) translation between the number of proposals and seconds is accurate, (4) these features are characteristic for the SA paradigm and do not occur all together in any of the other general optimization methods. Thus we have evidence that the brain mechanics cannot follow any of the other standard methods as well as evidence that SA is very close to the observed performance. Thus we conclude that *simulated annealing is the prime candidate for an ontological brain mechanics.*

7.7 Determining an Optimal Path from A to B

Suppose that you are currently at the point A and that you have computed that point B is the optimum that you would like to reach. In industrial reality, you cannot always just change all set points from A to B in one go. The process must be guided smoothly from here to there. This is called a *change at equilibrium* meaning that the change must happen at such a slow speed that the process is always (nearly) at equilibrium even though values are being changed. This will ensure that the process

---

problem instances contained in the TSPLIB collection of standard problem instances for both the TSP and CVRP. This collection represents the international testbed for TSP and CVRP algorithms.

One possible criticism for this is that measuring cost deviation from optimum skews the human performance because the subjects do not control length but the order of the cities. It has been shown in the context of the TSP that the deviation from one journey to another (in terms of Hamming distance – the number of different bits in a vector) approximately scales with the corresponding cost [45].

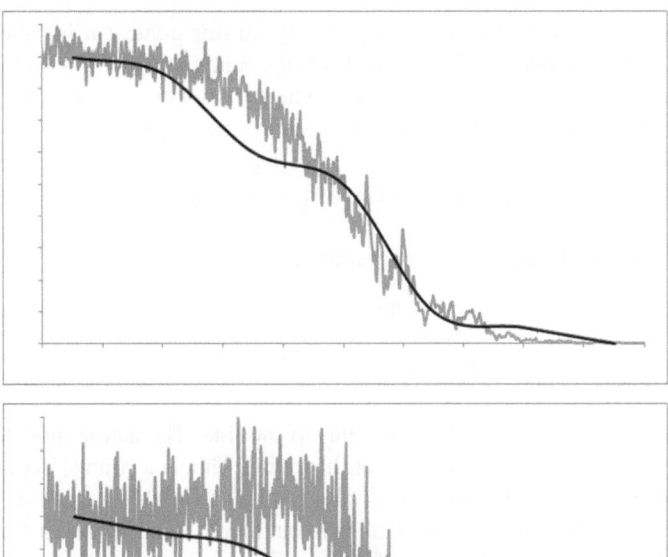

Fig. 7.3 The normalized cost deviation from optimum versus time in seconds is plotted in the upper image with the grey line being the SA output and the black line being the human subjects' average for a particular instance. The normalized standard deviation versus time is plotted in the lower image in the same manner.

will continue producing the product that you want without causing any unwanted side effects that may even destroy the optimization gains altogether.

This is equivalent to a navigation system in a car. It is unfortunately not possible to drive from *A* to *B* immediately but you need to pass through some of the intermediate points. There is an optimal route and it is the responsibility of the navigation system to tell what this optimal route is and to guide you through each of the steps involved as and when they are necessary and to alert you if you deviate from this plan. We must do the same for an industrial process.

A simple example of this process is to find the shortest line between two points. On a flat space, this is obviously a straight line. If the space is not flat (for example the hilly surface of the Earth), then this shortest path is no longer a straight line. The method of solving such problems is called the *calculus of variations*. This method has a few steps.

First, we define our criterion for optimality. In this context it is called the *Lagrangian* and is a function of the variables x, the function that we wish to find $f(x)$ and the derivative of the function we wish to find $f'(x)$, $L(x, f(x), f'(x))$. In the case of finding the shortest length distance between two points, we have

$$L(x, f(x), f'(x)) = \sqrt{1 + (f'(x))^2}.$$

Second, we define the *action integral* to be

$$\mathscr{A} = \int_A^B L(x, f(x), f'(x)) dx$$

where A and B are the two extremal points of the line. The action thus depends on the function $f(x)$, which is unknown at this point. This is a strange dependency as $f(x)$ is not a variable with a numerical value but it is a variable with a function as its value. We will not discuss this at length here but merely request the, in mathematics very common, "willing suspension of disbelief."

Third, we state that we wish $f(x)$ to take on that function as its value that will minimize the action. The result, after some pencil work, is the *Euler-Lagrange equation*

$$\frac{\partial L}{\partial f} - \frac{d}{dx}\frac{\partial L}{\partial f'} = 0.$$

This equation must be solved for $f(x)$. In general, this may be difficult. We will illustrate this with a simple example.

We start with the arc length

$$L(x, f(x), f'(x)) = \sqrt{1 + (f'(x))^2}$$

and observe that here

$$\frac{\partial L}{\partial f} = 0$$

as f does not appear explicitly in L. Then,

$$\frac{d}{dx}\frac{\partial L}{\partial f'} = \frac{d}{dx}\frac{2f'(x)}{\sqrt{1 + (f'(x))^2}}.$$

From the fact that this expression must equal zero, we see that we must have the numerator equal to zero and thus

$$\frac{d^2 f(x)}{dx^2} = 0.$$

The general solution to this is $f(x) = mx + b$, i.e. the famous straight line. This is the calculus of variations way of proving that the shortest distance between two points (on a flat space) is a straight line.

In industrial reality, we have multidimensional space (x is a vector) and we want to minimize the path's total cost in terms of the objective function that we used in optimization to find the optimal point. This will lead to an Euler-Lagrange equation that must be solved. As the objective function becomes the L used above, we cannot be more concrete than this here without a specific example. However, this equation will, in general, not be solvable directly but must be solved numerically.

That is how to determine the most economical path from A to B. The methods to numerically solve a non-linear second-order partial differential equation in several dimensions go beyond the scope of this book but may be obtained in several commercial software libraries for practical use.

7.8 Case Study: Optimization of the Müller-Rochow Synthesis of Silanes

Silanes are chemical compounds that are based on silicon and hydrogen. Important for industrial use, are the methyl chloride silanes. Industrially, these are principally produced using the Müller-Rochow Synthesis (MRS), which is the reaction, $2CH_3Cl + Si \rightarrow (CH_3)_2SiCl_2$. There are various hypotheses regarding the catalytic mechanism of the reaction but there is no generally accepted theory. Regardless of this, the reaction is widely used in large-scale industrial facilities to produce Di methyl chloride silanes $(CH_3)SiCl_2$ and Tri methyl chloride silanes $(CH_3)SiCl_3$, which we will refer to below as Di and Tri.

Practically, the reaction is carried out using silicon that is available in powder form in particle sizes between 45 and 250 μm with a purity of higher than 97%. The most common catalyst is copper and the promoters are a combination of zinc, tin, phosphorus and various other elements. The reaction is carried out at about $300°C$ and between 0.5 and 2 bars overpressure. In a fluidized bed reactor, the silicon powder encounters chloromethane gas from below. The product leaving the reactor contains the desired end product but also unused methyl chloride that has to be separated in a condenser. The mixture of a variety of silicones is now separated in a rectification where the desired Di and Tri are split off from the other methyl chloride silanes that are mostly waste. These desired end products can now be hydrolyzed into various silicones. The final products of this process can be practically used as lubricants in cars, creams for cosmetics, flexible rubber piping, paint for various applications, isolating paste for buildings and in a variety of other applications.

Unfortunately, the reaction produces several by-products that are unwanted. The selectivity of the process measures how much of the total end product is of the various types; for example a Di-selectivity of 80% indicates that 80% of the total end product is in the form of Di.

The market value of Di is highest among the different end products and so we want to maximize the Di-selectivity. However, the selectivity is influenced in part by the addition of catalyst. The relationship between increasing catalyst and increasing selectivity is a matter of folklore in this area. As part of this research project, no relationship whatsoever could be discovered within the range, 1% – 3% catalyst addition, studied.

As the catalysts represent a financial cost, the most economical selectivity is not, in fact, the maximum that could be chemically reached. We desire an economic maximum here.

Due to the fact that no generally accepted theory of catalytic mechanics exists, there is considerable debate and experimentation in an industrial setting on the correct use of the catalyst and promoters in order to get optimum performance. The question specifically is: In what circumstances should what amount of what element be added to the reaction? An important component of finding an answer to this question is what the desired outcome of adding these substances is. In an industrial setting, the commercial environment supplies us with some additional variables such as market prices and supply and demand variations. Finally, we establish that the desired outcome is a maximum of profitability. Whatever combination of catalysts, promoters and end products is required for this will be taken and it is the purpose of an optimization to compute this at any time.

Consider a black box. This box has five principal features:

1. There are various slots into which you feed raw materials such as silicon, copper and so on.
2. There are some pipes where the various end products come out of the box.
3. The box has a few dials and buttons with which you can act upon the system. These will be called the *controllable* variables, **c**.
4. The box has various gauges that display some information about the inside of the box such as various temperatures and pressures. These variables change in dependence upon the controllable ones but cannot directly be controlled and thus will be called *semi-controllable* variables, **x**.
5. The box also has gauges that display some information about the external world such as market prices for end products or the outside air temperature. As these variables are determined by the external world, we have no influence over them at all. These will therefore be called *uncontrollable* variables, **u**.

Inside this box, the Müller-Rochow synthesis is doing its job. Due to the lack of a theory about the synthesis, we cannot describe the process inside the box using a set of equations that we can write down from textbooks or first principles. Therefore, we will be adopting a different viewpoint.

Any industrial facility records the values measured by all the gauges and dials in an archive system that is capable of describing the state of the box over a long history. As the underlying chemistry has not changed over time, we therefore have a large collection of "input signals" (controllable) into the unknown process alongside their corresponding "output signals" (semi-controllable) in dependence upon the boundary conditions or constraints (uncontrollable), which, mathematically, are also a form of input signal. This experimental data should allow us to design a mathematical description of the process that would take the form of several coupled partial differential equations. Formally speaking, these equations look like $\mathbf{s} = f(\mathbf{c};\mathbf{u})$. Mathematically speaking, the uncontrollable variables assume the role of parameters (and hence follow the semi-colon in the notation) in this function.

Discovering this function is the principal purpose here and is very complex. One of the most intriguing features is that all three sets of variables are time-dependent and the process itself has a memory. Thus the output of the process now depends on the last few minutes of one variable and the last few hours of another. These memories of the process must be correctly modeled in order for this function to represent the process well enough to use it as a basis for decision making. In order not to clutter the mathematical notation, we will be skipping the dependence upon time that should really be added to every variable here. The modeling is done using the methods from section 6.5.

In order to do optimization, we need to define a goal g to maximize, which is a function of the process variables and parameters, $g = g(\mathbf{c},\mathbf{s};\mathbf{u})$. Using the recurrent neural network modeling approach, the goal becomes $g = g(\mathbf{c}, f(\mathbf{c};\mathbf{u});\mathbf{u})$, i.e. the goal is now a function of *only* the controllable *variables* and the uncontrollable *parameters*.

Optimization theory can be applied to this in order to find the optimal point $\hat{c}$ at which the goal function assumes a maximum, $g_{max} = g(\hat{c}, f(\hat{c};\mathbf{u});\mathbf{u})$. As the location of the optimal point is computed in dependence upon the goal function as described above, it becomes clear that the optimal point is, in fact, a function of the uncontrollable measurements, $\hat{c} = \hat{c}(\mathbf{u})$. Now we have the optimal point at any moment in time. We simply determine the uncontrollable measurements by observation and compute the optimal point that depends only upon these measurements.

Thus we arrive at our final destination: The correct operational response $\mathbf{r}$ at any moment in time is thus the difference between the current operational point $\mathbf{c}$ and the optimal controllable point $\hat{c}(\mathbf{u})$, i.e. $\mathbf{r} = \hat{c}(\mathbf{u}) - \mathbf{c}$. This response $\mathbf{r}$ is what we report to the control room personnel and we request them to implement. Ideally, the plant is already at the optimal point in which case the response $\mathbf{r}$ is the null vector and nothing needs to be done. As a result of the plant personnel performing the response $\mathbf{r}$, the optimal point will be attained and an increase of the goal function value will be observed; this increase is $\Delta g = g_{max} - g(\mathbf{c}, f(\mathbf{c};\mathbf{u});\mathbf{u})$, which we can easily compute and report as well. The relative (percentage) increase $\Delta g_{rel} = \Delta g / g(\mathbf{c}, f(\mathbf{c};\mathbf{u});\mathbf{u})$ has been found, in this example, to be approximately 6%, see below for details.

Please note carefully that the response $\mathbf{r} = \hat{c}(\mathbf{u}) - \mathbf{c}$ is a time-dependent response even though we have skipped this dependency in the notation. Thus, we do not necessarily proceed from the current point $\mathbf{c}$ to the optimal point $\hat{c}(\mathbf{u})$ in one step, see section 7.7 for a discussion of this point. Most often it is important to carefully negotiate the plant from the current to the optimal point and this journey may take a macroscopic amount of time – sometimes several hours.

Figure 7.4 displays this problem graphically using real data taken from the current process. The two axes on the horizontal plane indicate two controllable variables and the vertical axis displays the goal function. We can easily see that the

Current
Operational
Point

Sub-Optimal
Path

Sub-Optimal
Operational
point

Optimal
Operational
Point

Optimal Path

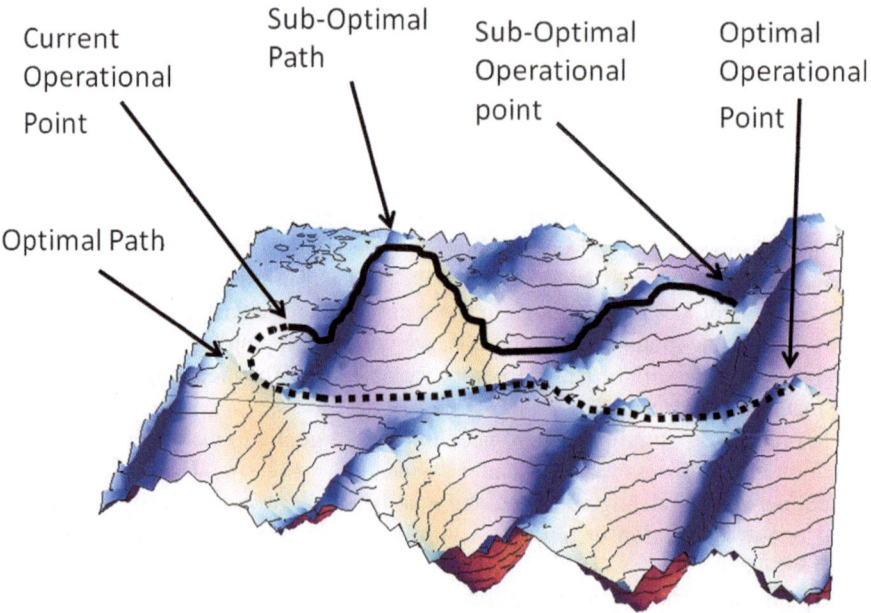

Fig. 7.4 The dependency of the goal on two controllable variables. The upper path displays the reaction of a human operator and the lower path displays the reaction of the optimization system. The paths are different and arrive at a different destination even though they started from the same initial state on the left of the image. The optimized path is better than the human determined path by approximately 3% as measured by the goal function.

change in a controllable variable can produce a dramatic change in the goal. The two paths displayed represent the reactions to the current situation by a human operator (the upper path) and the computer program (the lower path). They initially begin on the left at the current operational point. Because of their differing operational philosophies, the paths deviate and eventually arrive at different final states. This is a practical example of the human operator making decisions that he believes are best but that are, in fact, not the best possible.

For the specific current application, the molecules are produced in three separate reactors and then brought together for shipment. We are to optimize the global performance of the plant but are able to make changes for each reactor separately.

In this case, the controllable variables **c** were the following: Temperature of the reactor, amount of raw material to the jet mill, steam pressure to the jet mill, amount of Methylene Chloride (MeCl) to the reactor, pressure of the reactor and others relating to the processes before the synthesis itself.

The uncontrollable parameters **u** were: X-ray fluorescence spectroscopy measurements on 17 different elements in the reactor. The semi-controllable variables **s** are the other variables that are measured in the system. In total, there were almost 1000 variables measured at different cadences.

The goal function is the financial gain of the reaction. We compute the input raw materials and the output end products. Each amount is multiplied with the currently relevant financial cost or revenue. The final goal is thus the added value to the product provided by the synthesis. We desire this to assume a maximum. This function is dominated by two effects: Di is the most valuable end product and so we wish to maximize its selectivity and the overall yield represents the profit margin and so we wish to maximize it also. Possible conflicts between these criteria are resolved by their respective contributions to the overall financial goal. In the results, we will focus on these two factors.

The results reported here were obtained in an experimental period lasting three months and encompassing three reactors. The experiment was broken into three equal periods. During the *reference* period, the optimization was not used at all. During the *evaluation* period the optimization had only partial control in that the human operator controlled the input of catalyst and promoters. During the *usage* period, the optimization was given full control.

We may observe the results in figure 7.5. In each graph, the dotted line is the reference period, the dashed line is the evaluation period and the continuous line is the usage period. What is being displayed is the probability distribution function of the observed values. This way of presenting the results allows immediate statistical assessment of the result instead of presenting a time-series.

	Selectivity (%)	Yield (%)
Reference	79.8 ± 3.6	86.6 ± 4.2
Evaluation	79.9 ± 2.5	89.7 ± 4.3
Usage	82.7 ± 1.9	91.7 ± 3.2

Table 7.1 The results numerically displayed. For both selectivity and yield, we compute the mean $\pm$ the standard deviation for all three periods.

It is apparent, from the images alone, that we increase the selectivity and the yield with more use of the optimization and that we decrease the variance in both selectivity and yield as well. Numerically, the results are displayed in table 7.1. Decreasing the variance is desirable because it yields a more stable reaction over the long term and thus produces its output more uniformly over time.

We may conclude that the selectivity can be increased by approximately 2.9% and the yield by approximately 5.1% absolute. Together these two factors yield an increase in profitability of approximately 6% in the plant. We emphasize that this profitability increase of 6% has been made possible through a change of operator behavior only (as assisted by the computational optimization) and no capital expenditures were necessary.

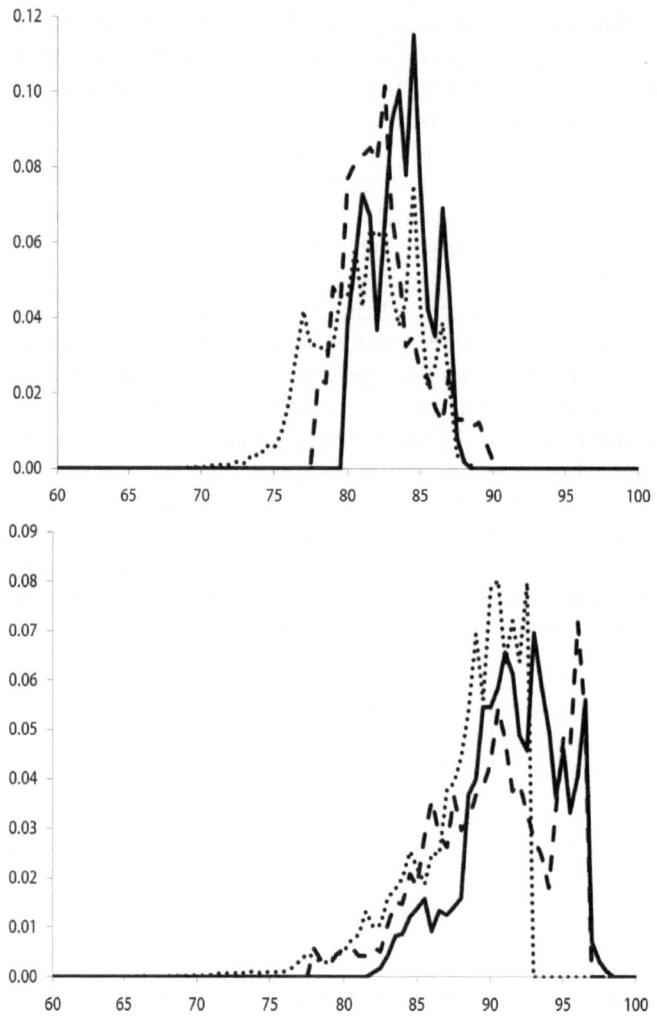

Fig. 7.5 The probability distribution functions for selectivity and yield of Di for periods in which the optimization was not used (dotted), used for controllable variables without the catalyst (dashed) and used fully without restrictions (solid).

7.9 Case Study: Increase of Oil Production Yield in Shallow-Water Offshore Oil Wells

Co-Authors: Prof. Chaodong Tan, China University of Petroleum
            Bailiang Liu, PetroChina Dagang Oilfield Company
            Jie Zhang, Yadan Petroleum Technology Co Ltd

Several shallow-water offshore oil wells are operated in the Dagang oilfield in China. We demonstrate that it is possible to create a mathematical model of the pumping operation using automated machine learning methods. The resulting differential equations represent the process well enough to be able to make two computations: (1) We may predict the status of the pumps up to four weeks in advance allowing preventative maintenance to be performed and thus availabilities to be increased and (2) we may compute in real-time what set-points should be changed so as to obtain the maximum yield output of the oilfield as a whole considering the numerous interdependencies and boundary conditions that exist. We conclude that a yield increase of approximately 5% is possible using these methods.

The Dagang oilfield lies in the Huanghua depression and is located in Dagang district of Tianjin. Its exploration covers twenty-five districts, cities and counties in Tianjin, Hebei and Shandong, including Dagang exploration area and Yoerdus basin in Xinjiang. The total exploration area in Dagang oilfield is 34,629 km^2, including 18,629 km^2 in the Dagang exploration area. For the present study, we will consider data for 5 oil-wells of a shallow water oil-rig in Dagang operated by PetroChina.

An offshore platform drills several wells into an oilfield and places a pump into each one. If the pressure of the oilfield is too low - as in this case - the platform must inject water into the well in order to push out the oil. Thus, the pump extracts a mixture of oil, water and gas. This is then separated on the platform. External elements like sand and rock pieces in this mixture cause abrasion and damage the equipment. When a pump fails, it must be repaired. Such a maintenance activity requires significantly less time if it can be planned as then the required spare parts and expert personnel can be procured and made available *before* the actual failure. If we wait until the failure happens, the amount of time that the well is out of operation is significantly longer. Thus, we would like to know several weeks in advance when a pump is going to fail.

Each pump can be influenced via two major control variables: the *choke-diameter* and the *frequency* of the pump. These parameters are currently controlled manually by the operators. Thus, the maximum possible yield of the rig depends largely on the decisions of the operators, defined by the knowledge and experience of the operator as well as the level of difficulty of any particular pump state. However, the employment of continuous and uniform knowledge and experience for the pump operation is not realistically possible as no one operator controls the plant over the long-term but usually only over a shift. Observation results show oscillations of parameters in a rough eight-hour pattern which supports the argument that a fluctuation in the knowledge and experience of human operators may lead to a fluctuation in the decision making and thus a varying influence on the operation of the rig. While some operators may be better than others, it is often not fully practical and/or possible to extract and structure the experience and knowledge of the best operators in such a fashion as to teach it to the others.

Pumps in an oilfield are not independent. Demanding a great load from one will cause the local pressure field to change and will make less oil available for neighboring pumps. Obtaining a maximum yield output, therefore, is not a simple matter but requires careful balancing of the entire field. In addition, certain external factors also influence the pressure of the oilfield, e.g. the tide. This high degree of complexity of the pump control problem presents an overwhelming challenge to the human mind to handle and the consequence is that suboptimal decisions are made.

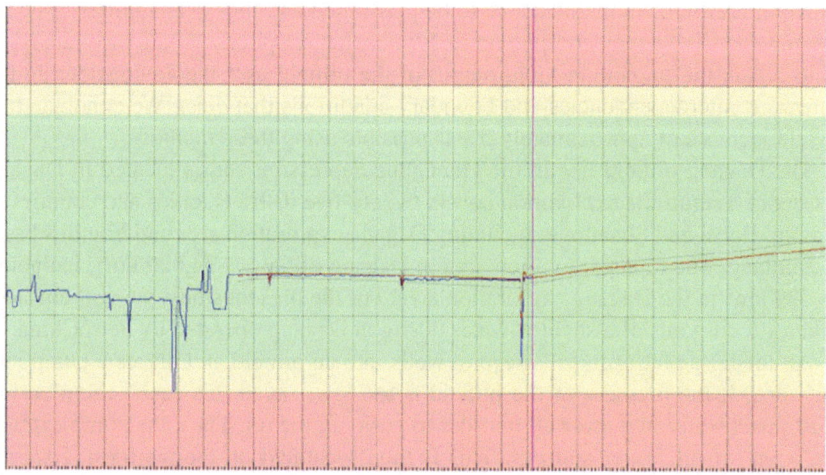

Fig. 7.6 The discharge pressure of a pump as measured (jagged curve) and calculated from the model (smooth curve). We observe that the model is able to correctly represent the pump as exemplified by this one variable.

The model is accurate and stable enough to be able to predict the future working of the pump up to four weeks in advance. It can thus reliably predict a failure of a pump for this time horizon due to some slow mechanism. We verify that the model accurately represents a pump's evolution in figure 7.6. The model was then inverted for optimization of yield. The computation was done for the entire history available of 2.5 years and it was found that the optimal point deviated from the actually achieved points by approximately 5% in absolute terms.

The main benefits of the current approach are: (1) processes all measured parameters from the rig in realtime, (2) encompasses all interactions between these parameters and their time evolution, (3) provides a uniform and sustainable operational strategy 24 hours per day and (4) achieves the optimal operational point and thus smoothes out variations in human operations.

Effectively the model represents a virtual oil rig that acts identically to the real one. The virtual one can thus act as a proxy on that we can dry run a variety of strategies and then port these to the real rig only if they are good. That is the basic principle of the approach. The novelty here is that we have demonstrated on a real rig, that it is possible to generate a representative and correct model based on machine learning of historical process data. This model is more accurate, all encompassing, more detailed, more robust and more applicable to the real rig than any human engineered model possibly could be.

The increase of approximately 5% in yield is significant as it will allow the operator to extract more oil in the same amount of time as before and thus represents an economic competitive advantage.

7.10 Case Study: Increase of coal burning efficiency in CHP power plant

Co-Author: Jörg-A. Czernitzky, Vattenfall Europe Wärme AG

The entire process of a combined-heat-and-power (CHP) coal-fired power plant from coal delivery to electricity and heat generation can be modeled using machine learning methods that generate a single set of equations that describe the entire plant. The plant has an efficiency that depends on how the plant is operated. While many smaller processes are automated using various technologies, the large scale processes are often controlled by human operators. The Vattenfall power plant Reuter-West in Berlin, Germany is largely automated in these parts. The maximum possible efficiency of the plant depends in part on the decisions of the operators, defined by the knowledge and experience of the operator as well as the level of difficulty of any particular plant state. However, the employment of continuous and uniform knowledge and experience for the plant operation is not realistically possible as no one operator controls the plant over the long-term but usually only over an eight-hour shift. Observation results show oscillations of parameters in a rough eight-hour pattern which indicates that a fluctuation in the knowledge and experience of human operators may lead to a fluctuation in the decision making and thus a varying influence on the operation of the plant. While some operators may be better than others, it is often not fully practical and/or possible to extract and structure the experience and knowledge of the best operators in such a fashion as to teach it to the others.

Furthermore, the plant outputs several thousand measurements at high cadence. At such frequency an operator cannot possibly keep track of even the most important of these at all times. This intensity combined with the high degree of complexity of the outputs presents an overwhelming challenge to the human mind to handle and the consequence is that suboptimal decisions are made.

Here, a novel method is suggested to achieve the best possible, i.e. optimal, efficiency at any moment in time, taking into account all outputs produced as well as their complex interconnections. This method yields a computed efficiency increase in the range of one percent. Moreover, this efficiency increase is available uniformly over time effectively increasing the base output capability of the plant or reducing the CO_2 emission of the plant per megawatt.

Initially, the machine learning algorithm was provided with no data. Then the points measured were presented to the algorithm one by one, starting with the first measured point. Slowly, the model learned more and more about the system and the quality of its representation improved. Once even the last measured point was presented to the algorithm, it was found that the model correctly represents the system. See section 6.5 for details on the method.

In the particular plant considered here, Reuter-West in Berlin, eight months and nearly 2000 measurement locations were selected as the history that was recorded at one value each every minute; yielding approximately 0.7 billion individual data points. After modeling, the accuracy of the function deviated from the real measured output by less than 0.1%. This indicates that the machine learning method is actually capable of finding a good model and also that the recurrent neural network is a good way of representing the model.

The power plant is largely automated and so we considered, for test purposes, only the district heating portion of the plant to be under the influence of the optimization program. The controllable variables would then be the flow rate, temperature and pressure of the of the district heating water at various stages during the production.

The boundary conditions or uncontrollable parameters are provided by the coal quality, the temperature, pressure and humidity of the outside air, the amount of power demanded from the plant, the temperature demanded for the district heating water in the district and the temperature of the cooling water at various points during the production.

The model was then inverted for optimization of plant efficiency. The computation was done for the entire history available and it was found that the optimal point deviated from the actually achieved points by 1.1% efficiency in absolute terms. This is a significant gain in coal purchase but mainly a reduction of the CO_2 emissions that save valuable emission certificates.

In the analysis, about 800 different operational conditions (in the eight month history) were identified that the operators would have to react to. This is not practical for the human operator. The model is capable of determining the current state of the plant, computing the optimal reaction to these conditions and communicating this optimal reaction to the operators. The operators then implement this suggestion and the plant efficiency is monitored. It is found that 1.1% efficiency increase can be achieved uniformly over the long term.

The model can provide this help continuously. As the plant changes, these changes are reflected in the data and the model learns this information continuously. Thus, the model is always current and can always deliver the optimal state.

In daily operations, this means that the operators are given advice whenever the model computes that the optimal point is different from the current point. The operators then have the responsibility to implement the decision or to veto it.

Specifically, an example situation may be that the outside air temperature changes during the day due to the sun rising. It could then be efficient to lower the pressure of district heating water by 0.3 bars. The program would make this suggestion and after the change is effected, the efficiency increase can be observed.

The main benefits to a power plant are: (1) processes all measured parameters from the plant in real-time, (2) encompasses all interactions between these parameters and their time evolution, (3) provides a uniform and sustainable operational strategy 24 hours per day and (4) achieves the optimal operational point and thus smooths out variations in human operations.

For those parts of the power plant that are already automated, the model is valuable also. Automation generally functions by humans programming a certain response curve into the controller. This curve is obtained by experience and is generally not optimal. The model can provide an optimal response curve. Based on this, the programming of the automation can be changed and the efficiency increases. The model is thus advantageous for both manual and automated parts.

7.11 Case Study: Reducing the Internal Power Demand of a Power Plant

Co-Author: Timo Zitt, RWE Power AG

A power plant uses up some of the electricity it produces for its own operations. In particular the pumps in the cooling system and the fans in the cooling tower use up significant amounts of electricity. It will increase the effective efficiency of the power plant if we can reduce this internal demand.

For the particular power plant in question here, we have six pumps (two pumps each with 1100, 200 and 55 kW of power demand) and eight fans with 200 kW each of power demand. The influence we have is to be able to switch on and off any of the pumps and fans as we please with the restriction that the power plant as a whole must be able to perform its intended function. A further restriction is introduced by allowing a pump to be switched on or off only if it has not been switched in the prior 15 minutes to prevent too frequent turning off and on.

Five factors define the boundary conditions of the plant: air pressure, air temperature, amount of available cooling water, power produced by each of two gas turbines. These factors are given at any moment in time and cannot be modified by the operator at all.

The definition of the boundary conditions is crucial for optimization. We recall the example of looking for the tallest mountain in a certain region. If the region is Europe, the answer is Mount Blanc and if the region is the world, then the answer is Mount Everest. In more detail, we have a set of points (the locations over the whole world) that consist of three values each: latitude, longitude and altitude. Out of these points, first select those matching the boundary conditions (Europe or the whole world) and then perform the search for the point of highest altitude.

In the power plant context, we must also define regions in which we will look for an optimum. We do this by providing each boundary condition dimension (the five above) with a range parameter. Let us take the example of air temperature. We will give it the range parameter of 2 degrees Celsius. If we measure an air temperature of 25°C, then we will interpret this to mean that we are allowed to look for an optimal point of the function among all those points that have an air temperature in the range

[23, 27]°C. As we have five dimensions of boundary conditions, we have to supply five such range parameters.

It is *a priori* unclear what value to give the range parameter. A typical choice is the standard deviation of the measurement over a long history. This gives the natural variation of this dimension over time. However we may artificially set it to be higher because the boundary condition may not be quite so restrictive for the application.

In the present case, we have chosen each range parameter to be one standard deviation over a long history. In addition, we have investigated a scenario where the range parameter of the air pressure is two standard deviations because we regard this to be less important.

In order to make our model, we have access to myriad other variables from within the plant. Thus we determine when we require which pumps and fans to be on in order to reliably run the power plant. This culminates in a recurrent neural network model of the plant, which we find to represent the plant to an accuracy better than 1%.

The model is now optimized using the simulated annealing approach to compute the minimal internal power demand at any one time. Operationally this means that the optimization would recommend the turning off of a pump or a fan from time to time and, aggregated over the long term, achieve a lower internal power demand by the plant.

The computation was made for the period of one year and it was found that the internal power usage can be reduced by between 6.8% and 9.2% absolute. The two values are due to the two different boundary condition setups. We therefore see that the loosening of restrictions has a significant effect on the potential of optimization. *Please observe the essential conclusion from this that the parametrization of the problem is very important indeed both for the quality and the sensibility of the optimization output.*

In conclusion, we observe that the internal demand can be reduced by a substantial margin (6.8 to 9,2%) which will increase the effective efficiency by about 0.05% to 0.06% given that the internal demand is only about 0.7% of the base output capacity. This is achieved only by turning off a few pumps and fans when they are unnecessary.

Chapter 8
The human aspect in sustainable change and innovation

Author: Andreas Ruff, Elkem Silicon Materials

8.1 Introduction

Imagine an everyday situation when driving to work except this time your usual route is blocked due to construction work for a couple of weeks. When realizing this would you not ruggedly wake from the brainless automation that literally makes you float to the office? And would you not quickly need to take back control over your turns in order to follow the detour? What distinguishes a rehearsed mental situation from the deliberate? How can humans actively participate in a change process? Finally, what is needed to sustain that change?

Continue to imagine that, when finally arriving at your desk, your office PC lets you know that your password expired. Most people - including myself - have a hard time to come up with a new set of letters, characters and numbers. Why is it so hard to think of something new and why do I struggle to use the new password for weeks? I enter the old password and only when it fails does my brain begin to question my action. Only then will my cognition remind me that I had to change it and that I have a new one. It will take a while until I get used to the new password. But after that it will become the usual password. The same applies to the example of the roadblock: Using the detour for a few weeks, you will sink into the same automation but with a changed route. It is the power of habit - the things we do regularly are processed in our brains automatically. A habit is acting on an accepted status quo and we tend not to think about it or even question its necessity or validity.

To strive for the new is given by our human nature. In order to alter any situation, it takes a lot more than just the intention to change. It requires the will and ability to take on the new status quo and live up to it. The complexity in remembering the password or the changed route to work lies in the conceived futility of the change and is enhanced by the users perception that this change does not make life easier nor does it contain additional value. In fact: The change is an obstacle.

Breaking the habit and accepting the new situation is hard both for someone initializing and managing the change and for the user. Overpowering habits and, especially, changing business procedures dramatically depend on every individuals ability to understand and accept the need for change and to adopt the new practice. In this text, I will discuss the various aspects concerning the human ability to change and its influence on sustainability.

In the past, I kept asking myself how much the sustainability in project work relates to a personal work style and how much is fortunate coincidence. I will summarize some important aspects on how to reduce the coincidence part. Additionally, I will mention some general aspects backed by research and literature. I want to share my personal experience and hope to illustrate the positive experiences that I made when dealing with people in project teams. I always used the human aspect in my project work and found that change against inner conviction is grueling and virtually impossible.

But how can human aspects help, especially when one is introducing major changes, and what can businesses do in order to make change sustainable? What preconditions are needed to make employees accept change or even help create it? I try to answer what successful change management is and how to gain sustainability. I will focus on the human perception of change and how the organizational set up and a managers personality can support the sustainability in change. All ideas, suggestions and examples derive from my experience in the chemical industry but I believe that they apply to any other business. My proposals and visions should be seen as a recommendation. I would like to give you ideas and food for thought on the value of the human aspect in order to increase your personal satisfaction and efficiency when making use of it. The terms sustainability and change are no contradiction. The status of any situation between two deliberate changes should ideally be sustainable.

8.1.1 Defining the items: idea, innovation, and change

Thomas Edison, the doyen of innovation and creativity, once said that the process of invention takes 10 seconds of inspiration and 10 years of transpiration. In order to understand the process of creativity better, a short differentiation of the terms used is useful.

The first step in a creative process is the idea. It can be described as what is before we think. It is a mental activity, the interaction of neurons and synapses resulting in an electric impulse comparable with a flash of light. This bioelectric interaction is able to create visions of objects or imagined solutions in our cognition. The brain is structured such that most ideas come up when the thinker focuses on something completely different or even sleeps. There are famous examples where scientists had worked intensively and concentrated to solve a problem but it was during relaxation that they imagined the solution. Kekule in 1885 worked on the structural form of benzene and dreamt about a snake chasing its tail. Researchers like McCarley (1982) affiliate this to the absence of the catecholamine adrenalin and noradrenalin resulting in reduced cortical arousal. The presence of catecholamine is related to the size of the neuronal networks and therefore depresses the individuals cognitive flexibility [74].

Emotional distance and a let-go attitude are very important when solving a problem. Ramon and Cajal (1999) mention this in their book Advise for a young investigator. Here they describe the "flower of truth, whose calyx usually opens after a long and profound sleep at dawn, in those placid hours of the morning that ... are especially favorable for discovery" [74]. Philosophers consider the ability to generate and understand ideas as the core aspect of individualism and the human being as a whole. Ideas are not isolated, they grow and improve when shared and discussed with others. The first idea is not necessarily the final solution, but it is the starting point of a thoughts long journey to realization. The bridge between the idea and realization is called creative innovation. The thought has to fall on fruitful soil or as Heilman calls it: the prepared mind [74]. It is obvious that the precondition for creativity is an open mindset and the ability to think outside conventional barriers. The prepared mind is able to look at the same situation from different angles. The open mind is not limited in its imagination and it keeps asking "what if?"

Kekule and his problem of structuring benzene can illustrate this concept. His dream about the snake chasing its own tail just induced the idea of a ring structure. The innovation lies in the acceptance of the idea and the insight that this is a possible solution. If Kekule would have discarded the idea of a ring shaped molecule because it just cannot be, he would have failed in solving the problem probably forever. The ability to "understand, develop and express in a systematic fashion" is the foundation of creative innovation [74]. Neither some special skills nor an increased IQ are required to be creative or innovative. Coming back to Edisons quote, the creative phase takes seconds but it may take years to establish the novelty successfully in practice. The goal of innovation is positive change to the status quo or simply to make something better. Innovation leading to increased productivity is the fundamental source of increasing wealth in an economy [129]. Innovation can be described as an act of succeeding to establish something new. Success in that respect means to eventually make an idea come alive. The novelty can be a service, product or organization. Once it is launched, there is no guarantee for success in the market. This can be seen when Amabile et al. (1996) propose [6]:

> All innovation begins with creative ideas. We define innovation as the successful implementation of creative ideas within an organization. In that view, creativity of individuals and teams are a starting point for innovation, the first is a necessary but not a sufficient condition for the second.

In order to be innovative one needs more than just a plain creative idea or insight. The insight must be put into action to make a genuine difference. For example, it could result in a new or altered business process within an organization or it could create or improve products, processes or services. Sometimes creative people have taken on existing ideas or concepts making use of individualization in combination with hard work and luck, making the replicated idea even more successful. Neither creativity nor an innovative mind can grant sustainable economic success.

In 1921, J. Walter Anderson together with Edgar Waldo Ingram founded White Castle in Wichita, Kansas. White Castle quickly became Americas first fast-food hamburger chain and satisfied customers with a standardized look, menu and service. In 1931, they had the idea to produce frozen hamburgers and they were the first to use advertisements to sell their burgers [132]. White Castles success inspired many imitators and so, in 1937, Brothers Richard and Maurice McDonald opened a drive-in in Arcadia, California. First selling hot dogs and orange juice, they quickly added hamburgers to the menu [130]. Today, White Castle sells more than 500 million burgers a year [131]. That sounds like a lot but compared to the market leader McDonalds, it is less than 1%. Surely White Castle was the first and is still in operation, but it is far from the success of others. The success of an innovation depends on several factors such as market conditions, customer demand and expectation. But more than that a good portion of luck is required: Being at the right place at the right time. That is what Edison meant when intimating that successful innovation is most of all hard work and transpiration.

This chapter is not about business plans nor will it advise the reader how to plan a successful business. I want to get an insight on what preconditions are needed to implement change successfully and, most of all, how to make it sustainable. The word "change" describes a transactional move from stage A to stage B. It is not necessarily true that the new stage is better than the original even if the intentions were good. A software update, for example, can put the user in a position of not knowing where to find buttons and features. The changed look leads to an uncomfortable feeling until the user gets to know the program better. This is similar to the above mentioned examples of the new password or the roadblock on the way to work. The software developers intention, of course, is to persuade the user with new tools and improved appearance. Change is perceived differently depending on the individuals situation. It is up to the user to experience the change as a chance or challenge.

Those of us who had the chance to manage change or work in teams in order to implement change have experienced that a substantial amount of money is spend to introduce advanced software, restructure organizations or improve processes or products. With all this cost and effort, how and why does the improvement slowly regress when the project team moves away? It is the human aspect that needs to be considered and remembered right from the starting point. It determines the sustainability in the change process. Enforced change is unlikely to be sustainable. As modern organizations need to be able to adapt quickly, the human aspect needs to play a central role in any change process.

8.1.2 Resistance to change

Most people are attracted by novelty. Especially when it comes to consumer electronics, thousands stroll over various trade shows or wait in front of retail stores to get the latest products. To be equipped with up-to-date fashion and to be trendy defines our status in modern societies. The speed of change is enormous. Companies underlie various constraints to keep up with the need to create new products and services at satisfactory prices. Challenges are to improve technology, market share or to stay (or become) competitive. The globalized world demands mental flexibility and the passion to take on any new situation. Yes, we grow with our responsibilities and we have to try hard to live up to the task. But too many businesses experience such pressure and expect more of the employees than they can handle. The modern (especially western) business world generates more and more mental diseases. The number one reason for sick leave is mental overload. The managerial challenge is to balance the need for change with their duty of care towards their employees. Businesses need continuous improvement and enduring change in order to survive and to continue engagement. Employees are required to contribute for the benefit of competitiveness, job safety and growth.

Studies done by Waddell and Sohal in the United Kingdom and Australia show that resistance is the biggest hurdle in the implementation of modern production management methods [126]. Interesting enough this resistance comes equally from the management and the workers. They also found out that most managers and business leaders perceive resistance negatively. Historically, resistance is seen as an expression of divergent opinions and good change management is often related to little or no resistance. That implies that well managed change generates no resistance. Here the question has to be answered what comes first, well managed change or to properly handle resistive forces. Enforced change, with the main focus on the technical aspects of change, seems to be widely accepted by management. Action plans are worked off and get reviewed but the individuals needs are often overlooked. It does not matter what is going to be changed, the reaction mechanisms of humans are similar and have to be taken into consideration. Resistance is a reaction to a transition from the known to the unknown [49].

Change is a natural attitude, but only if the initiation for this transition derives from the individual itself. If the initiation for alteration is pushed from the outside, then the change process needs to be well attuned. Every individual confronted with change undergoes the following phases: initial denial, resistance, gradual exploration and eventually commitment7. This is a well-developed natural habit of defending ourselves. Especially if companies execute on major business decisions (e.g. organizational rationalization by software implementation) resistance results from the individuals anticipated personal impact of the upcoming change. Humans immediately see themselves confronted with an uncertain future. Past experiences combined with what we have heard (from relatives who have been in a similar situation or the media reporting about others) escalate to existential fear and questions arise such as "will I lose my job?" and "do I have to sell my house?" Whether this anxiety is imaginary or real, the physiological response is that same: STRESS. The negative, irrational emotion represses any logical aspect affiliated with the intention and we divert all energy to defend the status quo rather than on the compulsory task. It is an ancient defense mechanism from deep inside that hinders us to consider change rationally, to adopt it or even to help shape it. Thus, resistance is often seen as an objection to change. But is that really the case? If resistance is negative how can we turn it into something useful to enable change? First of all, resistance and anxiety are important human factors for any undertaking. Data provided by Waddell and Sohal indicates that humans are not against change for the sake of being against it. In many observed cases, resistance occurs when those who resist simply lack the necessary understanding. Therefore, they have a negative expectation about the upcoming effect of the change. A major organizational restructuring or a local implementation of a software tool for process improvement will be seen (by some) as an assault. Preparatory measures should be used not only during the implementation but also as a structural tool.

It is important to realize and analyses resistance in order to make use of it. Managers and organizations should exploit resistance and use it with participative techniques. Participation in that respect means more than just giving regular status updates. Many companies inform their employees regularly about the status of an announced project. This is a one-way communication model and the problem is that, in many cases, questions of those involved are not or only inadequately answered. Bidirectional information and communication is a critical tool to create well-conceived solutions and to avoid misunderstandings or misinterpretations. After the intention to go through a change process is announced by top management, the middle management needs to communicate detailed information as soon as possible. Regular personal meetings with decision makers and their direct reports enable a consequent flow of (the same) information from the top down. Teams and departments should sit with their direct management and discuss actions, explain project milestones and intended results. The lower and middle management should provide platforms for addressing questions. This will keep up the communication even if the local leadership must not explain the exact plans and details.

Most of us have experienced situations where we lived through unconscious fear. It helps if one can express the sorrow and gets the sense that the fear is dealt with seriously. It is obvious that this cannot be done in a works meeting but has to be done in small groups or even in one-to-one conversations. Certainly this is a big effort and it (if done right) takes a lot of working time from the leadership and the work force but it will pay back when measuring the effectiveness of implementation. The organization needs to be set up to be ready for change. Thus, the organization should focus less on technical details but prepare psychologically and emotionally. It is mandatory for the management to define the goal and to set the expectations but it should also leave room for individualism and pluralism. The initial idea is not always the best. Leaders must learn to focus on the result rather than on every individual detail. The organization has to be constituted to articulate resistance and to deal with it. In return every individual needs to be confident that the managers are honestly willing to listen and communicate. Therefore the entire organization needs to be set up to ensure communication and a flow of ideas and suggestions no matter from which level of hierarchy they originate. This requires managers at any level of the entity to remember their vested duty to lead and manage people. The entitys organizational structure determines its ability to adapt to innovation.

8.2 Interface Management

8.2.1 The Deliberate Organization

Organizations depend on their ability to adapt to change and to make it sustainable. The following describes a desirable but fictional stage. To claim that all of the following is doable appears to be rather unrealistic. But I do claim that most companies have issues with regards to sustainable change management and that this is due to neglecting the human aspect. As little as resistance is seen positive in real companies it still is a desired and necessary process for sustainable change. Furthermore, the importance of a proper definition and scrutiny of functional interfaces is the key to this process.

Whenever peers have to work together the efficiency of this interaction depends on how well each of the individual knows what to do and how to do it. If there is an overlap of authority or improper definition of roles and responsibilities the involved employees spend a great deal of time and energy in organizing themselves. Some organizations even believe that employees will solve this conflict in the best interest of the business. The opposite is true. It is in the nature of humans to try to get the interesting or highly appreciated duties and to leave those that require a lot of work or are less esteemed to others. Instead of a cooperative atmosphere, a power struggle will eventually leave few winners and many defeated. Those who cannot or do not want to keep up with this fight might eventually leave the company. In addition to this, it is not necessarily true that the ones who emerge victorious are the better leaders or managers.

It is obvious that roles and responsibilities have to be defined. Reality proves that in many organizations grey zones exist. An organizational interface requires several aspects to be defined, executed and controlled. First of all it has to be defined which roles have to have a share in the interface and who controls and monitors it. Consider the following example from my experience when purchasing raw materials. In order to buy the right amount and to set up realistic delivery plans, sourcing needs to cooperate with manufacturing, quality control and perhaps even research and development (R&D) as well as logistics. Depending on the company and its set up, the finance department can be involved to contribute to payment terms, agree to letter of credit and to manage cash flow.

This is an obvious exercise and it sounds simple when taking the concept of value chains into consideration. If it is crucial to deliver high quality products in reasonable time at competitive cost, all stages in the value chain have to receive the right material at the right time. The smooth execution of customer orders requires everybody involved to focus on the same target: total customer satisfaction. Opposite to this concept, most companies have introduced individual Key Performance Indicators (KPIs) for every department separately. Who could blame the manufacturing manager to demand only top class raw material or the most reliable supplier if his task is to produce "just in time", reduce reject rate and keep inventory levels low. The manufacturing manager will simply not care for the cost of goods sourced because it plays absolutely no role in his list of performance indicators. Assuming the sourcing manager is held responsible for price reductions and availability of material, are these targets not conflicting? This fictional company will quickly find itself in the position that everybody is pulling the same rope, but in different directions. Who would win the battle, especially if personal bonus payments depend on the level of goal achievement? Furthermore, a proper interface definition also looks at the acting persons (defined as names or job functions) and defines who exactly has to take decisions. The principle of four eye decisions is expanded to six or more eyes. In the context of the procurement example, the manufacturing, R&D and quality leaders would all have to agree to the proposal made by sourcing. They could use evaluation forms and would discuss all pros and cons to finally sign off on the ultimate decision. Due to this, they share the risk of failure (instead of a single person taking the decision and being held responsible) and also improve communication. Throughout the process the team shares the need for transparent decision-making and traceable accountability.

This little example shows how important it is to manage any interface and to define all incoming parameter as well as the outcome specification. The better everybody understands the interfaces definition and the individuals share in it the more efficient it is. The critical definitions have to come from the management and it is within the responsibility of every executive leader to maintain the defined balance in each interface by frequently reviewing its overall performance. The benefit of well-defined, established and managed interfaces is that the need for change is easily detectable and its effect can be simply measured. Any diffuse organization with uncertain responsibilities unwillingly leads to mismanaged situations. The task of executing change is either hard to assign or it is given to someone who has to fight it through against the resistance of his colleagues. They will claim that the executing individual has neither the authority nor the power to implement any action. The battle for competence and power will make the implementation extremely challenging. The desired effect finally will fizzle out. The well-defined interface will remain well defined because the required change is assigned to the person in charge. Together with the necessary resources and helping hands of all involved, change can be implemented quickly and effectively.

A famous example is the work organization done by the Toyota Car Companys assembly teams. All work tasks are clearly regulated, described and monitored. Expectations are defined and the work outcome is permanently controlled. The individual is held responsible for his work and quality. If workers experience a problem, they pull a trigger and co-workers from previous and later work steps come together to discuss the issue, identify the root cause and the location of appearance as well as finding and implementing a solution. Later, they monitor if the implemented change is sufficient and track if the problem is solved for good. If they cannot find or agree on a solution, an upper hierarchy member has to be informed according to defined levels of escalation and will support the instant problem solving process or take other decisions.

8.2.2 The Healthy Organization

Let us assume that an enterprise defines, manages and controls its interfaces, all roles and all responsibilities. The resistance throughout the organization is still unmanaged and uncontrolled. Not knowing the reasons for change initiates resistance. In some cases this is aligned with the unwillingness of the leaders to exchange opinions. Maurer looks at resistance as a force that keeps us from attaching ourselves to every crazy idea that comes along. If an organization cultivates resistance and sensitizes itself for the human nature, it will provide better ideas and faster turn over in change. A necessary prerequisite is a certain type of manager, being able to change their attitude towards resistant employees and develop a healthy organization. This organization empowers managers to resist ideas originated in higher hierarchic levels. Every employee openly shares ideas and participates in a culture of discussion. A healthy organization is willing and empowered to improve initial ideas and shift opinions. It being healthy implies to have critical thinkers identified, accepted and involved at every level in the hierarchy. The appreciation of feedback and the willingness to shift opinion based on reasons needs to be practiced top down and should be lived as a business philosophy. Project teams must be put together based on both, their criticality and their professional experience. The more diversified a group is, the more facets are reflected and the more options are considered. It is critical for any environment to have strong characters with different opinions, whether it is a team, an enterprise organization or a group of peers. It is to the benefit of the leadership (at every level) to cultivate at least a few dissidents rather than surround oneself only with those who just say what one wants to hear. If resistance is used as a productive tool, it enables a positive environment of trust and honesty with an open dialog between all levels of the hierarchy. In such an environment alternatives can be considered carefully and thorough discussions can evaluate every option.

It is the duty of an organizations leadership to carefully select every member in the structure. Factors like specialist knowledge, leadership skills and personality play a major role when selecting the appropriate candidates. A healthy leadership has the guts to selectively pick those who are uneasy, unconventional and dare to speak up. It is in every managers hand to choose the appropriate team members – the healthy way is definitely stony and requires more effort from the beginning. I have chosen this way many times now and I have always been rewarded with excellent feedback, proactive flow of information and an extraordinary team spirit. All this contributes to a lot more than just the sake of implementing and sustaining change. It is of highest importance to make aware that a manager is not responsible for its direct reports only. The managerial duties also apply for those levels below, thus it is in every managers interest to have close communication beyond the direct reporting line. This ensures translation of business visions from one level to the other and their proper explanation. At the same time it guarantees everybody is and stays focused on that same target. In well functioning organizations the vision is clear and broken down so that the important parts get executed where needed. Many companies suffer over-communication. Visions, missions and updates are sent weekly if not daily and those who actually execute on these missions are not capable (due to IT access or language barriers) of understanding the message. Again, a good example of doing it right is Toyota. They have their targets visually broken down so that every employee can easily see, for example, how many cars need to leave the factory per day to make the monthly promise and they get their personal goals aligned.

Especially micro-managing organizations need to reconsider their position. As Antoine de Saint-Exupery states: "If you want to build a ship, dont herd people together to collect wood and dont assign them tasks and work, but rather teach them to long for the endless immensity of the sea." It cannot be overstated how important the information flow within an organization is. Sharing the vision means to provide the right parts of it and to transform the vision into executable and measurable tasks. A proper interpretation of the vision has to be communicated through the hierarchy with a clear and understandable obligation. From top down, managers have to sit down with their staff explaining the part of the vision they own and then break this part up into workable portions. These parts of the vision get assigned to the managers direct reports. The latter then do the same with their teams. Via this, the organization is truly focused on the common target and individual interpretation and/or cherry picking is barred. The proper break up and exact goal setting should be controlled over (at least) two hierarchic levels. A manager needs to set the goals for his direct reports and the managers superior controls that the set targets truly match the overall vision.

The vision itself is ideally defined long term. When breaking the vision down into manageable actions, take the time horizon into consideration. There are decisions made today with an immediate impact and there are actions paying off tomorrow. The leadership role of the top management is per definition oriented to the long term future. Visions of growth, EBITDA margin and profitability surely need to be transformed into actions with a time scale of months or years. This is the leading part of the organization. However, a worker on the assembly line needs to get the tasks in a more compressed time horizon. The vision of job safety, salary increases and promotion perspective can be achieved by every days performance. Executable tasks need to be laid out in hours or shifts. This is the part of the acting hierarchy and its targets need to be monitored on the short term. Led by the management, the vision gets executed by the shop floor personnel. The responsibility is spread equally over the hierarchy. One could not succeed without the other – leaders need actors and vice versa. The organization needs to be perceived like a Swiss watch: every cogwheel is equally important, no matter what size. The real difference is in the relative ratio of lead and act. This ratio varies and naturally it is at a 100% lead at the top of the hierarchic pyramid and at 100% act at the bottom. Draw an imaginary line through the hierarchy at the level where act and lead take the same share. Here is a certain, hierarchic level and anyone below that level is likely to be more on the workers side. This fictional border is extremely important. It is this hierarchic level that is the most important interface. It should be seen as the front line in communication and needs to be supported in an extraordinary way. Those employees are the ones that will receive their part of the vision more lead oriented and will need to pass it on with a strong focus on short-term execution (act). They need to transform the business vision into something understandable, no matter what language, cultural context or expectation. The front line in the organization is where the sorrow of team members gets shared. At the same time those working at this front line step in for the top managements decisions on a day-to-day basis. Those at the front line need to be embedded properly in the decision process or at least get well prepared and need to receive all information and training up front.

Imagine a shift leader on night shift being responsible for a handful of workers with different backgrounds and histories. Today, a certain change in the business strategy gets announced via e-mail by the management. This shift leader is the single point of contact that every shift member will turn to. If the shift leaders supervisor did not prepare him with pre-information or supplied him with answers to most likely questions, how can one believe that the shift will continue to focus on the job? Everyone on the shift will spend a great deal of time with recurring questions like "does this affect me and my family?" This time is costly and must be avoided. I am not claiming that the less sophisticated are not capable of comprehending financial data or business numbers – the opposite is true! But remember that a rational view on the uncertain new situation might be dismissed if the change affects oneself. It would be better to hand out individualized communication packages with background information and a few answers to likely questions. This information sent out shortly before or even with the announcement and to all relevant functions is necessary, especially in times of major business change. If done right, the abstract decision is explained and can be discussed. Concerns may be addressed and the focus will likely remain on the work rather than on something else. Depending on the impact of change, the team coming together has to be prepared well. Key personnel needs frequent training in communication and how to handle such an extraordinary situation. Most change does not come by surprise over night and communication can be prepared. The official announcement is the top of the iceberg only. Preparing the key information multipliers with required information is mandatory for sustainable change. Town hall meetings with all employees are good to maintain communication regarding the change process but only a few will have the courage to express personal fears or raise questions in public. It is best to break up the enlarged meeting into work teams and further explain the situation in smaller groups.

This chapter is not about leadership or professional management and there are many other articles, publications and books available. Nevertheless, one aspect seems of importance to me when discussing communication and organizational responsibility. It is a question about leadership itself: how many employees can possibly be led effectively by one person? In times when efficiency and productivity are the main drivers for business decisions, many (especially larger) companies trend to merge departments and divisions. In order to make the organization financially more efficient, groups and independent teams are put together under just one leader. The number of direct and indirect reports keeps growing continuously.

To me, the maximum number of direct reports should not exceed 5-7. Following the thought path of the effective vision communication, any manager is liable for at least the second hierarchic level below himself. For example, a manager with five direct reports would need to intensively work with all 30 (25 + 5) individuals. If one takes the leadership role seriously, this consumes a great amount of the daily working time. On the other hand, the fewer hierarchic levels exist, the more day to day business inevitably will end up on the managers desk. At a certain point the leadership time gets repressed by the excessive time that functional involvement requires. To me, that stage is reached in the majority of companies and it is an organizational disaster. It is not my intention to judge but to alert. Any organization decides how much a manager is a people leader or a functional worker. But my deepest belief is that anything above 40% functional work will be at the cost of effective leadership and good communication.

Time is the resource of today and especially of the future. It will not take long until (especially) highly educated professionals will ask for more balance between job hours and private time thus for a better work-life balance. There is and furthermore will be a trend towards flexible and effective working with new work structures such as home office, split of labor, etc. Time already is but certainly will become a more satisfying form of payment rather than pure monetary salary. Modern organizations will have to react to the fact that more managers (male and female) value family time over career. Any organizations challenge for the upcoming years will be to make use of the existing resources as effectively as possible. The Lean principle with its concept of labor efficiency is not able to model this by far.

People management and modern leadership will have to focus much more on the individuals involvement in e.g. teamwork, projects or cross-functional problem solving. Thus the manager shall have more time to identify and promote talents and, at the same time, encourage the one being less motivated. Dealing with the latter is extremely time consuming but necessary, if taken seriously. Assuming a manager takes the time and spends it on a one-to-one conversation with a critical individual. Then the individuals supervisor needs to be coached and instructed as well. All this requires at least three peoples work time. The alternative is to do nothing but this consequently will de-motivate the entire group. Assuming that there is a fixed number of less motivated people in every organization, enlarged teams are not necessarily easier to manage especially as hierarchic levels might have been eliminated. Following the idea of sustainable change, the new manager will have to focus on all individuals, ultimately identifying their talents and level of motivation, working through communication and structural issues and finally gain new team spirit. How can this person possibly find the time to work on the newly assigned job tasks?!

This conflict gets even worse if those who have the most professional qualification are promoted to manage the merged department and not the ones with excellent leadership skills. The number of professional and leadership tasks suddenly overloads the new manager and there is a chance for the individual to get sick or leave the company. The worst case is if the manager decides to set priorities towards less people management. Thus, the frustration in the team increases with similar consequences for the employees health and the fluctuation rate but multiplied by the number of reports. Leadership and people management cannot be learned as easily as professional skills. The higher the rank in the hierarchy the less important professional skills are and the more important leadership and emotional intelligence become.

Healthy organizations consist of smaller highly specialized teams supervised by those who continuously prove to be motivators and enablers. The base for the healthy organization is the self-motivated worker who shares the managements vision and sees a clear perspective for himself. The workers talents need to be discovered and his resources should be managed properly. It is the managerial task of any supervisor or leader to bring the best out of their teams. The hierarchic pyramid needs to be turned upside down putting the worker and its individualized work environment in focus. Those directly involved in the value chain need to be supported most. The above-described front line is what needs the highest attention, as this is where the managerial vision is practically put into action. Here is the interface that makes businesses successful. The entire remaining organization is supportive only. Empowering the organization means to develop manageable structures with clearly defined interfaces, responsibilities and authorities. A healthy organization is equivalent to hierarchic emancipation.

8.3 Innovation Management

One issue of highest importance to long-term business existence is consistent success. Both rely on continues innovation. As shown above, innovation is not being the first but being better than the others. Google started with an innovative algorithm to search the Internet but they did not invent the search function itself. They came into business by advancing existing technology. Only later in their history, they introduced revolutionary technology like Google earth, the virtual street view or providing an online free counterpart of Microsofts office software package. Furthermore, they were the second (after Apple) to develop their own smart phone. Innovation is Googles backbone and each employee must spend 20time on non-job related issues. Googles success is based on diversified employees and the trust that the freedom of thought will result in creative ideas. As much as I support this concept, it is hardly applicable for those in traditional business with non-virtual products and rather high manpower cost. Those who operate in the real world need more basic tools to generate and implement innovation to be successful. The recently introduced industrial operational excellence programs use Lean, Six Sigma and other problem solving techniques. Lean relates to the Toyota Production System and Six Sigma is a statistical program, which was originated by Motorola in the 1960s. Both systems are common answers to the same problem: structural operational improvement. In many organizations the two are implemented together.

Lean focuses strongly on the executing workforce and its ability to prevent failure immediately. In Japan, detecting defects is not viewed negatively or to phrase it differently: The Japanese are controlling the process to detect the issue as soon as possible. This, in combination with a culture being subservient to authority, worked well for Toyota. In Japan, no one would change work procedures or dare to question the importance of details. Only if there is a problem, the belt is stopped. In the event of a shut down, the team joins and discusses the issue. Together they find a solution and implement corrective actions immediately. If the issue is out of the teams authority or cannot be fixed with a certain, well-defined time frame, the problem gets escalated to the next hierarchic level. That level is equipped with more authority and might have the power to either empower the team or call for help. Continuing the escalation until the problem is solved guarantees that eventually something will be done about it. The belt stands until the solution is implemented. It is a common aim to continue the belt as quickly as possible but also to strive for perfection even if that binds the forces.

This is opposite to most western civilizations were individuals tend to hide or cover their mistakes. We tend to believe that errors will either not matter or someone else will fix it later anyways. The culture is fundamentally different, especially in the individuals identification with the product. While most western employees work for money, the Japanese work culture is close to a second family. I am neither glorifying the one nor criticizing the other – it is important to understand that a program is based on culture and cannot easily be copied to an organization with a different culture. The individualized western civilization requires corrections to the program. Pulling Lean, as practiced in Japan, over a central European work environment will fail. There are multiple examples of companies being successful with Lean and Six Sigma. Managing innovation is doable with the various programs and systems. In many companies, the traditional organizational structure has been adjusted to match the newly introduced improvement programs. Departments with specialists and expert know-how are introduced and, in many cases, the talented employees have been moved from elsewhere in the hierarchy. The introduced program creates localized innovation while the entire surrounding organization is reacting to the input. Hot spots of innovation and creativity are not enough to inspire an entire business. The innovation must come from within the organization itself and must derive from the deepest wish, the one in every human, to create and innovate. Everyone should have the same opportunity to put innovations and improvements into practice. Exclusive programs will not succeed in sustaining the change they implement.

Everybody is innovative. Inventions are as old as mankind and it is not surprising that people even today continue to invent. Thus, making use of ideas and developing creative solutions has become more vital for economic organizations than ever. The preconditions for innovation are adequately discussed, but the question remains how to bring the best ideas forward and how can the organizational structures support it.

Operational Excellence continues its triumph in almost every enterprise. The larger businesses allow themselves the luxury to allocate resources in new departments or structures to implement and execute the systematic improvement. The smaller companies introduce those excellence systems within their existing structures. Lean, Six Sigma, Production Systems and innovation programs are introduced to ensure implementation of innovation management and systematic improvement. Black Belts are put almost everywhere with the clear objective to discover and eliminate waste. Statistical tools are installed, trainings held and the company policy gets adjusted to the new philosophy. Consequently, there is a lot of change in the organization: new faces are employed with new-sophisticated titles (many of them in English) to execute on those fancily named programs. Imagine being a worker at shop floor level, facing the structural shift and discovering the situation. You would be challenged with a double impact: (1) Improvement projects will unforeseeably change your work environment and (2) you will have to accommodate to a diversified organization with additional reporting and responsibility structures. The individuals position becomes more diffuse and established organizational structures change. This particular change is driven top down and is littered with terms that most workers do not comprehend.

The acting participants are different from those who have to live with the alteration. How can one believe this will work without friction or at least with some resistance? The situation gets even worse when facing the fact that most improvement leaders (e.g. Black Belts) average retention time is a little over two years. The duration of stay is influenced by the projects size, its importance and the companies attitude towards its own program. Would it not be silly to promote that project manager who has just familiarized himself with the team, figured out shift structures or even came close to the problems root cause? Would you not agree that it could easily take up to two years to discover all this? I confess to be rather critical with the programs and its improvement leaders. For over 100 years, the industry in central Europe did not know these highly specialized and centralized functions. To me this is a consequence of the extreme reduction of manpower. Those working in an area for a long time and being highly experienced did not need intensive statistics or other tools to discover fundamental issues.

I am not claiming that analytical (especially statistical) know-how can be replaced by experience, but I strongly suggest combining it. The young, inexperienced professional with his special knowledge and his technical ability is in a much better position when being introduced (not only to peers and colleagues) by an older employee who is well respected and familiar with the situation. I predict faster and more sustainable results if the generations work together and combine their individual strengths. Team success strongly depends on the characters in the game. Successfully implementing improvement starts when preparing the organization for the improvement program itself. Thus, organizational development depends on the ability to think critically and the respect for age and experience. Neither can the program replace the experience, nor can we implement profound effects without modern analytical tools. Sustainable process change and significant improvement cannot be prescribed. Many things need to fall into place to sustain change. If statistical figures, graphs and key metrics are presented to the management, the viewers need to fully comprehend what is shown to them and they need to grasp what consequence might come along with the improvement.

The biggest issue with improvement projects (especially of complex processes) is that an improvement here can have a major impact elsewhere. Amendments should be real and not a statistical fake or an imaginary effect on colorful slides. As soon as the number of improvements or the speed of its implementation enables career opportunities, the chance for improper project management with short-term effects increases. The individual promotion should be linked to the improved quantity, but mainly to the projects quality. Responsibly caring program managers will always ask for the voice of the customer when evaluating a projects success. The customer in my eyes is, by definition, the one that needs to operate and maintain the improvement later on. This is unlikely to be a manager or supervisor, but mainly and certainly the person dealing with the change afterwards. It only takes moments to judge the quality of the implementation, the documentation and the users involvement during planning, execution and training. In a healthy organization, poorly managed change is identified immediately. Consequently, it gets attacked and stopped.

Criticality and resistance are important factors and they prevent us from making stupid mistakes or repeating the same mistake. Healthy, in that respect, also implies that the management takes the resistance seriously and values it. At the same time, communication is open and everyone strives for necessary change. A healthy organization will be able to adopt any trend quickly but, at the same time, will wisely adapt it to meet its needs.

8.4 Handling the Human Aspect

As outlined earlier, there are multiple aspects were the human interface impacts sustainable change. Many of the described company-wide introduced programs contain change management within the improvement project phases. Unfortunately, neither of them really looks into the human sensitivities within the change. It is not part of university education plans to prepare their students for the upcoming challenges nor is it included in management or leadership seminars. Any new leader faces this fundamental issue when being promoted into a leadership position. Together with the new professional task comes the responsibility to lead a team. The challenge of leading is accompanied by the struggle to compete with colleagues and other departments. The issue with this overload lies in poorly defined roles and responsibilities and improper preparation of the individual itself. The ignorance of proper leadership is carried from one level to the other and sooner or later the entire organization lacks of human interface management. Some leaders do learn from experience. Throughout our professional lives we experience managers with positive as well as negative leadership attitudes. It is important to remember that nobody is perfect. It is of highest importance to take on some of the positive abilities presented by managers and peers.

The principle for successful lead-management is "tread others like you want to be treated yourself." Good and pure people management is not easy and lacking time for leadership induces most interpersonal issues in companies. Effective tools to manage the human aspect in sustainable change are wishful. Their effectiveness strongly depends on the overall companys attitude towards this aspect and every individuals intentions. Each of the following topics is an important factor by itself.

But, when combining them into a leadership vision and living up to it, it will increase engagement, motivation, identification and, finally, productivity and revenue. Focus on the human aspect is not only a matter of sustainable change but also a chance for sustainable business success. The following suggestions are not ranked nor are they a guarantee for success. They are meant to add some practicability to the paragraphs above.

8.4.1 Communication

Team-Meetings should be held (at least bi-monthly) with a clear, pre-communicated agenda and an open space for critique and suggestions. The team should be selected by function such as shift leaders or regional marketing managers. On recurring events, an extended team, including e.g. the deputy shift leaders or the most experienced marketing manager, should get an opportunity to be heard. They also need to have their share in hot topic discussions and must have a chance to express their opinions. Especially in times of change, listening frequently to those affected can reduce tension and resistance. Remember that those who resist are not necessarily against the change. Change pressed through by tension and force is likely to increase the resistance. Team meetings are excellent tools to develop a common sense and to communicate the status quo. Sharing the information with the team enables a broad discussion and allows everybody to participate. The major disadvantage of multi-people meetings is that those who gain a major share in the discussion are those who would have probably expressed their opinion anyways. You need to get to those who are quiet and meet them in personal communications.

One-to-one meetings should be held frequently with the leaders but also with those known to be the unofficial leaders. Communicate and involve them during decision-making and have special focus on those who struggle accepting the change. Special attention is needed for those being extremely reticent. They are likely to not be brave enough to speak up while among others. But, you need to find a way get to their viewpoint, too. An open atmosphere in a non-business environment (cafeteria or a colleagues office) will help to establish this channel of communication. The individual needs to be and feel safe to disclose his opinion or express deep sorrows. Follow up meetings are important to double check that the person is still okay and on track but also to clarify any misunderstanding or misinterpretation upfront. Oneto-one conversations are extremely time consuming but indispensable. I have made great positive experiences when demanding critique from my direct reports. That way, you force them to think about what needs to be changed. Or to phrase it slightly differently, the question to ask is: "what do I have to change in order to make you more successful?" One-to-one meetings can be planned as an official meeting but could also be set-up as a consultation-hour on fixed dates. Getting to the crucial point of information is difficult when dealing with people. You never know whether you are being told the truth or if your counterpart just plays politics. A good feel for people is the most essential characteristic of any leader but if you really need a good average opinion, you should use anonymous communication tools.

The Critique Box can be a letterbox or a web-based anonymous message drop box. Most, but especially international, companies have compliance hotlines and ombudspersons. Here, people can address their concerns and questions and receive help. This rather complex system might even prevent people from using it. Users could get the impression that their concern just is not important enough to bother such a professional instrument. And if they place the concern, will it be treated seriously and confidentially? Certainly people do not want to feel or be disadvantaged when using the hotlines. An anonymous web-based interface is a simple tool and every employee can submit inquiries. Quick and frequent feedback is most important if the company takes it seriously and deals with the requests. Handling the inquiries and dealing with whatever comes in is the challenge the leadership has to take on. There could be a team consisting of the general management and the representatives of the workers union to process the incoming information. Depending on the type of feedback, the project manager and key users could provide facts to ensure a satisfying answer. In any case, both the question and the answer should be published closely to where the input came from. Place them for example on the intranet and hang them on dedicated information boards for everybody to read. Others might have had similar questions. The attitude and honesty of the answers is extremely important. The answer must be reasonable and comprehensible to the questioner. Write as clear as possible and try to really answer the question. The tool is extremely powerful if used properly. When providing reliable communication, and if the sender is treated seriously and respectfully, confidence in the (project) management will certainly increase.

Reliable communication or "do not say it if you do not mean it" is an obvious suggestion. It sounds easy but it is not. We all know that political statements are sometimes hard to believe. Take the criticality of the national finance situation as an example. The deficit in most countries is so dramatic that it constrains the room for any political manoeuvre. A statement in which someone promises tax reduction may feel implausible. Nevertheless, certain politicians are able to get attention although their main statement seems illusory. Whoever communicates, wants to transmit a message. The key is the combination of the speeches context and the speakers body language. A plausible combination of both may determine the receivers emotion. Some people think about communication as a pure exchange of information. A very good example to prove that this is wrong is President Barrack Obama. He was not yet elected and already many people put great hope in him. This is mainly because he got contemplated as a new type of politician. He received the Nobel Peace Price after just a few months in office because he managed to transport the hope for a better world. It is important to communicate with a positive attitude and to transmit an optimistic message. This is of highest importance especially in official announcements. You want to spread your optimism amongst the audience and would want them to look positively into the future as well. It is obvious that the right words are needed but, more than that, you need to believe in them yourself! Many are skeptical about what they hear or what is promised to them because they have been disappointed many times before. Promises never came true and so many projects that were sold too well got stuck half way. Honesty and reliability are the foundation for the trust and cooperation needed to create sustainable change.

8.4.2 KPIs for team engagement

Once a project is started and the team is selected you will find that there is an inner project circle that will be driving the initiative. The so-called "core team" needs to be supported by experts who were not chosen to join it. The latter is called the outer circle. Both groups need to be engaged in the change activity and every single player must stay focused, committed and satisfied. Frequent checkups on the common understanding of the project's base (target, approach and timing) enable you to effectively manage the project and to quickly overcome identified hurdles. You need to make your teams emotional involvement measurable and visible. This is a leading indicator for the projects, and your personal, progress and success. Communication either in team meetings or on a one-to-one base is vital for the project management but hard to quantify.

Questionnaires do help to get a quick overview of where the team thinks it stands. Is your opinion on progress aligned with the teams view? Or is your perception impaired? Modern online forms help you to design questionnaires within a few minutes. Even the evaluation is simple. Just select the receivers click a few times and you are done. Unfortunately, there are a few more things to consider. In order to make the evaluation quick and easy, I suggest avoiding free text fields unless you want participants to comment on a special topic. You should use predefined statements instead and continue to ask the user for his level of agreement. The grading for your statements could be 1-5 where 1 represents the least consensus and 5 represents highest. Alternatively you could go for 0-100% or chose any other way. A few examples for precise statements are

1. The project is on track and we will finish by the end of the month
2. Communication is reliable
3. Project management takes on suggestions

If you do this assessment frequently you can detect disagreement early and take action if needed. Focus on binary, precise and clear questions. Look at the examples mentioned above. Asking whether communication is reliable and frequent might put the user into a dilemma to answer both questions at the same time. What if it is frequent but unreliable?! Also remember whom you ask and prepare separate questionnaires for individual groups if necessary. Example: in the preparation for a major enterprise resource planning system (e.g. SAP or Oracle) implementation, a group of people is asked to rank how much the system is perceived to simplify their daily work. Assume the result to be rather confusing as about 50% view the system as a useful tool and the others do not. Furthermore imagine that the more technical oriented group would see less use then the rest of the group. Would you not simply split the total into separate user groups to get a more diversified answer? Tracking the number of supporters within the technical group could also be an excellent KPI for the commitment of the originally more skeptical user group. In addition you have easily identified what group you need to focus on in order to increase the total engagement and commitment.

The happiness check is a very simple way of monitoring the emotional baseline. If you take 30 minutes and stand in the hall, walk over the site and look into offices, just count the number of happy faces. The "smiles rate" is no solid scientific indicator nor should you overvalue it. But, and especially if you start somewhere new or launch a project in an unfamiliar area, you can get a feel for the current morale. Frustration and resignation are sub-optimal conditions for launching a change project. If you find the latter to be the case you should focus to identify the cause of the dissatisfaction first and then try to use it in your favor. Develop an emotional leverage to get people out of their lethargy. Convince them to help you to change their situation. Encourage them to be part of the creative force rather than to continue complaining. The happiness check can also be used to measure the attitude towards you and your project. Again, do not overvalue it and also consider the environmental circumstances. I would expect more smiling, happy faces during spring and summer compared to autumn or winter.

Visualization and non-verbal communication enable you to communicate to teams even without being present. Especially when interacting with the outer project circle visualization is the tool of choice. Try to put as much as possible into graphs and pictures. Replace text blocks by bullet points and use short but precise wording. All communication has to be profound and reliable. Avoid speculation or guessing and, if you have to, separate your guesses and mark them clearly. If visualization is used as a standard communication tool it reduces subjective perception. Print it as large as possible (e.g. poster size) and hang it in a highly frequented place such as the entrance hall, cafeteria, waiting area or the smokers room. Start to pick random people and present them your project by using the poster. Ask what they think and if they agree with your statements. Even if the volunteers are not impacted by the project nor involved in it, they might raise excellent questions or give you hints on misleading information. Thus you get the chance to inform people and spread the information you want to be spread. In addition, you may get detailed feedback free of charge. Put a feedback e-mail-address or a telephone number on the poster so that you can be contacted for questions or suggestions later on. Keep the poster as a record, especially for the project documentation. In order to make your poster easily comprehensible use dedicated areas for certain topics. Place a project summary preferably at the pages top or bottom and provide a status indication for the various sub-projects. Color codes or status bars can be used to indicate its progress. Colors as used in traffic lights are recognized as green equaling "OK", yellow as "behind" and red stands for "critical." A status bar is more detailed and contains additional time related information. You could use weeks or months, milestones or the number of items worked off. Simply calculate the percentage and illustrate it in a bar chart similar to what most people know when downloading or installing computer software. Alternatively or in addition, you could place a little marker over the status bar to visualize where you are supposed to be in accordance to plan. Add a text box next to your status bar to point out the reasons for a delay and articulate required actions and help needed to speed the project up. Especially if your project is behind schedule, you need to think about proper communication to get the support you need. The earlier you indicate items as "behind" or "critical" the higher the chance to counteract.

A first year review is done to analyze the status of the project after one year. Many projects, even if they get planned and executed well, lack after-project support. Unfortunately, many project resources are cut back or responsible leaders are assigned to new tasks before the ultimate implementation. As a result, the users are quickly in charge to finalize the project and they handle all the problems. What if those who are suddenly in charge were least involved in the project? It is a managerial culture to ensure that expectations are met! Everybody who is buying a car will test drive it before signing the contract. In professional life somehow, the executive attention ends with the announcement of the projects end. As soon as the first few results are reported, some managers believe in its continuation and, in many cases, focus on the next projects. Sustainable project management requires regular meetings to follow up on the status. This includes involving the users to ensure that the expectations really are met. Expectation in that respect can vary depending on the function. Unfortunately the preferred project expectation is whether the spending is within budget. Functionality or performance as promised is second in line. To me, the primary expectation in any change is to sustain the desired new state. Change of operational behavior and the related improvements depend on repetition. Be aware that learning can erode over time. There is no such thing as a quick hit. Behavior practiced for years will not change in a month. Sustainability equals change for, and especially over, a long time.

8.4.3 Project Preparation and Set Up

Stakeholders and team members play a critical role in any project. Team members are everybody working on the change process. By definition, a project team consists of multiple professions and departments. In many cases, the team consists of employees that do project work besides their everyday jobs. Most companies have been through rationalization programs and it can be that each position only has one employee. There are no spare capacities that can be pulled in full time to work on projects. Only if there is a major investment to be launched, experts might be made available. But even then, a team member might be part of other teams as well. This person with his individual character and profession might play a more or less central role in each team. Thus, an individual might be critical for the overall success of multiple projects.

In distinction to a team member, the stakeholders enable the change. Stakeholders are not part of the project team but supervise the activities. Project managers need to communicate with key stakeholders frequently. They demand updates and supply help to overcome hurdles during the project execution. Take the following example: Imagine a launch of a global marketing project. Delegates from the regional marketing organizations gather as the project team. The global product manager would act as the project leader driving the initiative and being supported by the team. Stakeholders could be e.g. the global marketing manager and/or the global sales manager. It looks like the set-up is simple and roles and responsibilities are clearly defined. Expectations seem to be set and frequent communication is ensured. This project should be straightforward, successful and sustainable. Reality proves that most of the projects (especially the ones driven globally) are a debacle. The simple but devastating cause is that too many projects need to be handled by only a few team members. In addition, the individual often is left alone when it comes to balance project time and their everyday duties. And even if the individual would be dedicated to full time project work, those that ultimately execute it are probably not.

Back to the example of the global product launch as described above: Imagine that the regional marketing managers would be dedicated to that project only. What are their duties? They need to work with the regional marketing groups to prepare the promotion. They need to make sure that production is aware and ready for local manufacturing. Quality standards need to be discussed and agreed upon globally. In order to be effective and unique, they commonly need to be accepted locally. There are various local interfaces to work with in order to coordinate and control the information flow to and from all involved departments. If just one interface is not managed properly or fails, the entire project is at risk. This is not an issue if one assumes all parties concerned are pulling the same rope. But what if conflicting targets are real? Do all the important projects that get launched almost daily really take into consideration that project members do work in more than just one team? What if a potential team member simply cannot be freed from daily duties? Would you agree, that everybody is pulling the same rope but in different directions? Just like the business vision, the projects need to be assigned top down and resources need to be planned and allocated reasonably. Most project managers tend to forget that there is more than what happens within their project. They forget that any action needs to be executed by someone in the outer project team. The executing individual probably does not get allocated nor planned. Regrettably, the executing forces likely set the pace of the implementation and finally the likelihood of sustainability. I claim that the number of projects within businesses is far too high to be managed properly. Or to phrase it differently: Fewer projects with proper resource allocation (at all levels) will increase the chance for successful and sustainable change.

The Team Selection is an important act for any project. You should identify two or three well experienced supporters, even if the task seems easy and the implementation is perceived as a no-brainer. Assigning a team does not necessarily mean to hold meetings and sit for hours regularly. Any idea must grow and the condition will only remain changed if someone feels responsible and takes care after the project is done. The earlier in the project, the end users are involved and the closer you keep contact with the key personnel, the more sustainable it will be. So why not build a small team and spread the responsibility as well as share the credit? Another important aspect is the potentially reduced resistance. The lone fighter, even with the smartest idea and a brilliant brain, is likely to fail. Selecting the right (trusted and respected) team members helps to knock down prejudices and helps you around roadblocks. And remember that it is not always easy to get the users' honest opinion. Most people would rather express their concerns to a colleague than to a project manager. In order to be supported, you need to be perceived as helpful. Allies are needed to get to that stage quickly. The sooner you establish an honest working routine with the (informal) area leaders, the faster and more successful (and sustainable) your project will be. Focus on frequent communication and you might get all relevant information without even asking for it. Remember to deal with resistance openly and try to find the root causes rather than to fight it with force. Keep an eye on your project teams attitude towards the change. Once your team does not believe in success, consider re-adjusting the project rather than the team!

Delegation needs to be learned. Delegation is putting trust and confidence in your associates. The delegate represents the department and must have full support. Make sure you chose the right occasion to send delegates. Official meetings with the general management or the works council are likely to be un-delegable. Imagine the effect on the attendees and the delegate if he could not answer questions or even contradicted himself. Politically, you need to stand up for your project yourself. For any other issue, a delegate can be assigned. Some project managers keep complaining to be overworked and stressed. As much as I sympathize with that, I also question whether they do well in delegation. As stated above, choosing the appropriate team is the baseline for success and a good work-life-balance. Select the team members who are capable and assign tasks to them. Make sure the delegate understands the goal and is not overwhelmed by it. Delegated tasks still need to be supported. Offer your help frequently and request updates. Monitoring and controlling the project still is the project managers responsibility. Delegation done wisely can help to reduce the managers involvement in sub-topics and ensures his ability to see the big picture. Proper delegation will get things off your desk and the receiver might even see it as an opportunity or chance. Most people that I work with are grateful to participate in project work. They like to get away from the daily routine and to get different insights. They get motivated when having a say in business decisions. Trust and esteem put into the delegates will energize them and will enable them to deliver highest performance. Be aware that delegates are speaking in their managers name and that they represent the entire project. Their decision should be binding and must not be questioned or revised unless absolutely inevitable.

8.4.4 Risk Management

Risk and Opportunity evaluation is a necessary act in managing businesses. Some modern companies do not evaluate well and go for any new idea, especially if it comes from the upper levels in the hierarchy. A natural diversification between locations, culture and regional requirements gets abandoned in favor of organizational simplicity. Dealing globally and being successful in local markets requires diversity. No doubt, markets are different but demand is local. When the US car market still requested high horsepower vehicles, the trend in Europe and Asia was already going towards less emissions and city friendly mobility. Products and especially brand marketing are focused on localized needs. Most companies fail to make us of the different cultural strengths. Just think about Toyotas production concept. It works fine in Japan but needs corrections in order to work in Europe or North America. Business Unit structures covering continents and countries need to manage the diversity properly in order to benefit from it. Opportunities can be straightforward in one country but have high risks in another. Local regulations and laws can be different in the various regions. To implement solutions blindly, just because they worked elsewhere, is likely to create problems. Thus, look at the potential solution first. Try to understand what the possible consequences are when implementing the solution. Do you really have a similar issue that needs to be solved and what might the result be if one implements a solution for a non-existing problem? Solutions are powerful and finally sustainable only if adapted reasonably. Global guidelines and programs should be made variable. They need to be flexible to comply with global policy once they have been adjusted to every region. Risks and opportunities can be judged best locally. Nevertheless, there is a need to control the entire process and to set up a supra-regional organization.

Risk management structures need to be considered carefully and shall be maintained in any organization. If a business decides to introduce operational excellence or similar initiatives, the various options for setting up the organization need to be evaluated. Depending on these structures, the working attitude as well as the approach to manage risks and opportunities will be fundamentally different. One option is to centralize all global experts in one department in order to send them out as in-house improvement consultants. This concepts obvious advantage lies in a unified approach and standardized tools being used. The solution transparency is high as all information about local projects are tracked and documented centrally. Best practices and benchmarks get collected, evaluated and get looped back into the organization. There is only a small risk that every-day duties overwhelm these resources. A separate improvement organization ensures that improvement leaders focus 100% on the task. It can even be of advantage if the improvement expert is unfamiliar with the area. Too much expertise in a particular area contains the risk of routine-blindness. Similar to consultancy from an external company, the projects duration is critical and thus the total success depends on the involvement and commitment of the executing teams (outer project circle).

An alternative set up could be to empower and enforce existing structures and resources. Here the local managers and their teams get trained to enable the improvement by themselves. The advantage of this set up is to have a direct (reporting) line to the process experts. An improvement manager based locally might instantly know what to change and whom to contact in order to get things done. By giving them the right tools you might enable the change directly. Local people have a strong network and it enables them to get to the cause quicker than non-locals. Furthermore, the non-local might be perceived and treated as an outsider even if employed by the same company. Getting things done often depends more on interpersonal relations than on organizational power. A well-known program leader, who ideally operates within a strong emotional network, might be able to reduce resistance towards change dramatically. That is exactly where the biggest risks reside: As much as the personal relationship enables quick execution that much does it prevent necessary cuts and/or drastic decisions.

I had the pleasure to work with a highly knowledgeable process manager who had over 40 years of experience in the area. He told me to not ask for his opinion about a change in the process. He expressed to know the existing process for too long to be able to imagine how it could be done differently. He suggested splitting duties: I would tell him what I would want changed and he would ensure execution, if doable. The combination of specific process experience and program know-how worked well in that case. Incorporating the program approach globally and executing it locally is the biggest challenge when designing corporate improvement programs. Who leads? Who decides? And finally who executes? It is a question about core competences, roles and responsibilities and finally power. Once the dispute "global capability versus local competence" and "program know-how versus process experience" is solved, the improvement work may begin.

The steering committee controls the overall project progress and direction. It approves if milestones got worked off and keeps track of required next actions. Some project managers conceive steering committee meetings as an offense to their competence and professionalism. I consider steering committees helpful and important if it comes to responsibility sharing. Once the committee approved a milestone, it acknowledges your effort and you have a regular platform to raise questions and concerns.Working with steering committees is easier if you have prepared an excellent meeting agenda based on the projects schedule and planning.

First, you need to collectively agree on the level of detail in the project plan. It depends very much on the initiative if a complete schedule for each individual action is required or if it would be sufficient to assign completion dates to milestones.

Second, set up a meeting and discuss your project plan with the committee as early as possible and also agree on the KPIs to monitor the projects progress. Once you start executing the plan without the committees improvement it is hard to redo the fundamental planning. This applies to major investments in civil engineering or construction and to smaller projects. Spend sufficient time on the planning and ensure frequent and timely communication with the steering committee especially in the projects initial stage.

Third, the steering group should also discuss if alternative plans and fallback positions are required in the event of obstacles in the master plan. You and the committee should agree and include a written procedure to your project plan with proper definitions about who needs to be informed in the event of complication or failure. In addition you should plan upfront which additional resources might be pulled in or can be requested in such a case. The definition of a crisis, as well as planning for its management, is done best if prepared timely. Some project managers are not aware that project delay or failure can easily cause major business issues. Mismanaged projects have the potential to harm the company not only by the pure fact of misspent investment money. Apart from a massive impact on the businesss image (e.g. the incident on the deep water horizon caused at BP in 2010) there can be legal liabilities. Imagine being responsible for a major capacity expansion project. A delay might have substantial impact on customers material availability and therefore contains high risks of contractual penalty. Would you not agree that it might make sense to share this responsibility? In order to do so, every level of escalation as well as the communication paths needs to be defined and approved by the steering committee beforehand.

8.4.5 Roles and responsibilities

Sustainable Responsibility is one of the key factors in sustainable change. In order to sustain change it is critical to keep up the force that ultimately initiated the alteration process. There is a difference between doing the right things and doing the things right. This is a fundamental challenge. The residence time of the new state is determined not only by organizational structures but by the managerial attitude towards project work in general. Assume that a reengineering project in the chemical industry is launched. A team of highly professional project managers and technical experts is put together and all of them are dedicated to this task only. A budget is available, steering group meetings were held and the targets and timelines are approved. Everything looks in order and the project seems to be driven towards sustainability. The dilemma lies in different personal expectations that may contradict sustainability. What if the project leader got promised a bonus if the budget gets under-spent by some percentage? Would you agree that his personal ambition and expectation is set already? What if the leading project manager would know upfront that he is to be promoted to manage the department after the project? Would he do it right or do the right things?! Doing it right is not necessarily sustainable. Is it right if the project is within schedule and budget and minimum requirements are just met?

Doing it right could also involve doing the right things and eventually even over-spend budget if that would ensure advanced technology leading to a more reliable process. I am not proposing that one is better than the other. As a project manager you need to understand what the expectations are and as a business you need to know what you want and how to achieve it. You should deal openly with the expectation and communicate it accordingly. Albert Einstein once said that "we cannot solve problems by using the same kind of thinking we used when we created them." Change, if well thought through, should ideally solve certain problems or improve the status quo. Well thought through in that respect means not to accept that issues may arise elsewhere. Relocating problems is not an option in sustainable change management. Sustainable responsibility also implies that the responsibility remains with the leader even after the projects closure. Coming back to the example of the reengineering project: Would you not agree that the project manager acts responsibly if supplying spare parts for the newly installed equipment? The areas maintenance budget should not be in charge to fix project issues once the budget and project team are gone. Sustainable change needs holistic responsibility. Bedouins move from one place to another and leave, when they are gone, little to no impact to the environment. I was able to observe that same behavior by global engineering teams. They moved on to another project and left little improvement but caused a great stir.

Inventor driven change is the most natural form to implement modifications. It lies in human nature to create, invent and improve. Historically, humans have always sustained the new status quo when they have seen a clear advantage compared to the old state. The ideal contender to implement the change is the one who had the idea. Practically speaking, you should enable the inventor to take the lead to implement his own idea. Who could explain the underlying problem and convince colleagues or co-workers better than the solutions originator? The inventor is motivated to ultimately solve the issue. Anyone else might give up after a few attempts. The innovator who believes in his idea will continue to try. In a professional situation, the amount of work spent and the total number of attempts certainly need to be restricted and controlled carefully. My unfortunate but practical observation is that too many ideas never get implemented. Innovators do not share their ideas as they fear loss of intellectual property. Many of them believe that their idea needs to be ready for implementation before disclosing. Many companies have suggestion tools where employees input their ideas and get rewarded if their suggestion is put into practice. It is a pity that once the raw idea is submitted, it will be evaluated and implemented (if at all) by someone else.

The originator has extremely little influence on the process and his colleagues and peers often do not even know that the idea exists. Imagine a brain pool of ideas were everyone could browse through, get inspired and add input and thought to others suggestions. It would be like an open community where the order of ideas and innovations is tracked carefully. Everyone adding reasonable input gains the same share of the final idea or suggestion. The originator is nothing without those adding practicality and vice versa. You could take your standard suggestion form and add a couple of text boxes to the back. Document the basic idea on the front. Describe the suggestion in as much detail as possible and name the originator, date and time. Publish it within the working community and if someone wants to add to this idea, they use one of the boxes on the back. Thus, the idea gets considered carefully and refined. It is important to not jump on the idea immediately. Let it sit and age like a good red wine.

Imagine the inventors could be asked to lead their own idea into realization. What if you could enable them with resources and help them make their idea come to life. The minority of the proposers do it only for the money. I state that an idea, realized by its inventor(s), per se comes with a longer sustainability. The implementation is done with brain and heart and the inventor(s) will do the right things right. But even more important is that the ideas fathers are known. This project will be perceived as change coming from the inside of the organization. Improvement being introduced by outsiders often is viewed as imposed. Suddenly the change gets a name – indeed name the equipment or improvement after the inventor and put a label next to it (if doable) in memory of the event. The emotional relation to the improvement made by someone known is different. Although the outcome of the improvement obviously is not any different, the handling and the sustainability are. It is the same emotional differentiation as if you drive a rental car or borrow it from a friend. In both cases the car is not yours but you might treat them quite differently.

Self-discipline is a necessary precondition for any behavioral-based change. Discipline can be ordered and controlled but once it derives from within a person, it is a lot more powerful. Let self-discipline grow by naming the innovator, by teaching the intentions underlying principles and by presenting the expected benefits. You should never underestimate the users. If they lack the discipline, do not try to force change on them as the change will fail and a lot of resources will be wasted. Discipline needs managerial leadership and increases in an atmosphere of trust and honesty. Work on the people first and try to create a team spirit. Without the discipline, do not implement change.

I was once assigned to a process improvement project in a chemical synthesis area. Rather than starting to implement the necessary technical solutions I tried to harmonize the way the shifts ran the process. I established frequent meetings to discuss the various operational philosophies when running the process. A common understanding of what the process is, what it needs and what actions should be taken set the baseline. This discussion took months until we came to a workable agenda. The common understanding resulted in a shared commitment towards standardized behavior. This is not necessarily self-discipline yet but the resulting peer pressure forces discipline. The impact of a shared commitment towards harmonized operations reduces the induced variability of human interaction. Once the impact is positive and obvious, the discipline will follow naturally but slowly.

The human aspect is great and powerful no matter whether the improvement is technical or organizational. And it has two sides. Support will help you to manage the situation easily but destructive forces destroy even the best idea or manager. It is extremely hard to fight your ideas through against negative confrontation. In that case, you need subversive activities. Try to find the root cause for the rejection and identify those willing to support you. Start with those being positive and start in tiny steps to prove your concept right. If you fail, do not give up. Always go back to the entire team to discuss the results, adjust the trial and commonly agree to start over again. Make sure to work on the teams self-discipline by involving them in the decision process and keep on selling the advantage that is in there for them.

8.4.6 Career development and sustainable change

Stagnation and Sustainability are two totally different things. Stagnation is sustainable but not vice versa. Stagnation maintains the status quo. A good indicator for stagnating organizations is when people state, "it is OK the way it is" because "it has always been that way". Stagnating organizations are inflexible and unimaginative. As shown above, flexibility in mind and a creative vision are drivers for innovation. Organizations need to understand that the ability to accommodate new situations is a personal strength. Human Resource departments should identify those candidates being open minded and those who see the opportunities rather than the problems.

Creative employees need to be developed and deserve the chance to witness the organization's flexibility and opportunities. Employers need to make use of every individuals strength independent of educational level and degree subject.

Talent Development in that respect requires accepting and managing risk. Imagine a successful employee working in an almost perfect position. Would it not be unfair not to promote this employee just because the organization has no successor for that position? Or, would it not be sad if that same employee misses out to apply for a job just because there might be no option to go back? I am not in favor of massive rotation within organizations but I deplore that too many creative heads are stranded within inflexible structures. Simply there are too many underdeveloped talents. This major problem gets worse as the population ages and soon it will get even harder to recruit qualified staff. The problem is self-imposed as the resources are cut back and some organizations simply lack a minimum number of people. The human resources departments are degraded to count heads rather than develop talents and increase the resource efficiency. Sustainable improvement is a tool to release some of those talents that will initiate improvement somewhere else. Here lies a fundamental opportunity for businesses. Consider all the in-house experts already working in your structure. It is just a matter of will and a bit of clever investment in the right resources. Managing successful and sustainable change is not necessarily restricted to employees with university exams.

Proper Resource planning takes all the above-mentioned into consideration. It is critical to look ahead and prepare the organization for the future. If an experienced employee retires, the knowledge is gone forever. Specialist know-how cannot be transferred to a successor in weeks; it might take years. Identifying the ideal candidate and training this person on the job are necessary requirements for sustainability. Success relates not only to know-how or experience, it often relates to the employees personal and professional network. Building up such a network is stony and slow. Once established, it helps to generate ideas and reduces the risk of repeating mistakes over and over again. The successor needs to be trained properly. That includes the time needed to interconnect and to build up the network within the organization. The training also needs to incorporate the professional preparation for potential leadership roles. Leadership is gained by experience, not by taking classes. The experienced leadership skills are even more important if the leader is not the manager and still needs to get things done. Cogency is more powerful than persuasiveness.

In order to plan and track a projects resources properly I use mind maps to catch, sort and rank ideas. I prepare individualized action lists and assign them to people in a timely manner to be able to control their work. According to the slogan "how to eat an elephant – bite by bite," I define sub-projects and assign resources to them. I define the minimum requirements and assign a sub-project leader. Despite the need for planning, you should not overdo it. Try to balance complexity and clarity. This can be done when one master plan shows an overview and multiple sub-plans and action lists are created for each sub-project, month or person. The level of detail differs with individual preferences and the project volume. Make sure to discuss the plan and its timetable frequently with the team to control the status and to ensure the teams awareness and commitment to the common goal. Even if you use computerized project planning tools, print the plan for the discussion and post it somewhere publicly so that everybody can see it. The more open and transparent the project is, the easier it is for the project team to stay focused and for the outer project circle to comprehend and actively participate.

Best to the top needs to be more than a phrase. It enables career opportunities for those being willing and capable. It is unacceptable to not invest in the human asset. The career of talented candidates needs to be supported and coached by professionals. A transparent program with support at every level of the organization ensures a constant talent flow. This implies that the lower hierarchic levels need more attention as their number is higher and their (average) age is lower. Thus, this group contains more potential to be explored and developed. The senior leader who made it from the lowest level in the organization is emotionally bound to the company. He brings expertise, experience and a working network that outsiders probably do not have. External options should be used only if the outsider really is better than the internal solution – true to the motto: best to the top.

Award and Recognition systems are supporting tools to career development programs. Not everybody can follow a career path but everybody wants to be recognized and awarded for performance. The award and recognition system is not necessarily monetary in nature. It has to be set up to enable managers and leaders to access it immediately. Employees need instant performance response but not only when things went wrong.

Such a system could start with the allowance to buy ice cream during hot summer days or grant teams to order pizza by themselves. Celebrating success empowers the entire organization to be proud of performance. The system promotes the performer but also e.g. provides loans to selected talents when building a house. This monetary system focuses on employee retention and the double positive effect for the employer is obvious:

1. The employee is emotionally bound to the company and benefits from a low interest loan. It is unlikely that this employee will leave the company during the payback time. Once owning property, the immobility increases resulting in strong regional inflexibility. That gets especially important if the area is rather unattractive to work in (e.g. little industry or reduced availability of jobs).
2. While the employer uses the tool to motivate identified individuals, he might even make some money from the interest, depending on the company size. This money could be used to finance training opportunities.

Certainly, there is a risk of privileging individuals and there will be discussions about the systems fairness. Thus, I stress the need to limit the value of the immediate recognition and rather have frequent team events (sports, dinner, cultural events, etc.) and recognize individuals with an award but without money. Those awards could be presented in a funny way and should not only be limited to work performance. The employee of the month could be accompanied by the "we are glad you are back" award for someone having been sick for a while. It is less the award but more the recognition and the teams expression to really be glad that the individual is back. This also creates emotional binding, which is so important for employee retention. Another attempt could take the family situation into consideration. Why not give "time" as an award? One could be rewarded with a half working day and an entry voucher to a local swimming pool in order to take the kids for a swim. The award lies in additional quality time with the kids. The joy and fun will be remembered and might continue to have a positive effect on the employees motivation.

8.4.7 Sustainability in Training and Learning

Culture of failure seems like a revolutionary concept. It implies that more learning experience derives from failure than from success. If failure is analyzed properly and the right conclusions are drawn, then this statement is true. In most cases, we do not question why something is working – only if it fails do we start to realize how things work. After Thomas Edison invented the light bulb, he stated that he had found a way to make it work. In addition, he learned over 500 ways how to not make it work. His innovations are based on a learning experience discovered from failures. Conclusions from results are driving forces in innovation and are not limited to technical processes only. For example, realizing that the information flow within an organization is bad opens opportunities to restructure and gain potential business advantages. Improvement is possible only when detecting the issue and consequently analyzing and reporting it. A culture is needed where nonconformity, failure and defects are seen as a chance. Employees need to be safe to discover, alert and admit mistakes. It is a managerial responsibility to protect the information source. The organization should award employees for detecting failure and the general culture should be to deal openly with mistakes. Focus on solving the problem, e.g. like Toyota does: Bundle all available forces to improve the process.

Train to change and practice sustainability. Everything needs to be learned and practiced, even the systematic of change. In a professional environment, improvement is often linked to rationalism and economization and is therefore mentally connected to job cuts and perceived injustice. The training sessions should clarify the mission of the improvement process. Most companies provide tool training and teach how to manage change when introducing improvement programs. The training should also teach how to think out of the box and how to gain creative ideas. There are several tools available to experience the creative thought process. A lesson could start with a small example to demonstrate that some things are not impossible even if they seem to be.

Take the following example: draw a box and in it 9 circles (3x3). Ask your colleagues to connect all of these circles with just three straight lines. The task seems impossible. The clue is to remove imaginary boundaries, see figure 8.1.

1. There is no rule that overdrawing the box is prohibited.
2. There is no rule that the circles must be connected at the center.

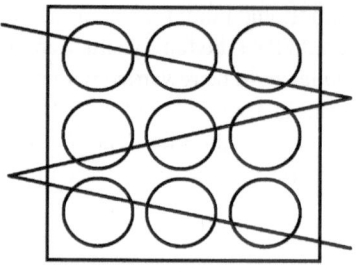

Fig. 8.1 The solution of the puzzle in the text.

This little exercise could be a starting point. You could continue to show example projects from within your company and explain how change is managed and how the results affected the work environment. Most companies do not intend to cut jobs but if it is to do so, make it clear upfront. If you want the change to be sustainable, the rules must be clear and fair. Cutting jobs in one area does not necessarily mean that people get laid off over the whole enterprise. Will the affected persons be used to fill open positions elsewhere in the organization or is there even a chance to continue to work as an improvement leader instead? Rationalization may free up resources desperately needed somewhere else in the company (please also refer to career development).

Diversification is a source of different ways to solutions and a surprising learning experience. As much as the overall business goals need to be defined unambiguously, the way to achieve that goal has to remain flexible. Too many boundaries will limit the creativity of the executing forces. Every problem has more than just one solution. Different departments, locations or business units have different backgrounds and individual needs. The teams have to have the freedom to discover their own ways and to develop their individual solutions. This will assure a more focused approach and increase commitment, as this is perceived as "our" solution. In addition, it broadens the experience baseline and fills the toolbox. Adapt the standard tools and adopt individualized ones. Every situation is unique and consequently individual tools need to be developed to design the matching solution. This may take longer and probably require more resources but it is a fundamental precondition for sustainable success.

8.4.8 The Economic Factor in Sustainable Innovation

Employer attractiveness is an increasingly vital fact if companies want to maintain their business. The struggle for the best talents and various employee-binding strategies were described earlier. Business success in that respect depends on, but at the same time generates attractiveness to those considered as "the best". Porsche is still one of the most desired employers in Europe. It is the brand that sells but also attracts young, ambitious talents. Forming an employer brand is more important than anything. The chance to be part of a highly innovative and well-respected brand is a major employee binder by itself. Choosing from the best is a luxury that only very few, huge industrial enterprises had in the past. In the last few years companies like eBay, Google or Facebook became extremely attractive to work for. It is the opportunity to bring ideas to life that attracts especially young people. Sustainable innovation and development are critical factors in developing existing employees as well as attracting the highly talented new ones. Another remarkable factor can be discovered when comparing the Internet based companies success to those dealing in traditional industries. I refer to e.g. the automobile or the consumer electronics industry as the traditional industry. Even the banking and insurance sector can be named as traditional although they got very much virtualized in the last years. The major difference between the two is the product application. Innovation and quick releases of new updates as well as an innovative website are the drivers for web based companies. The quality in innovation is perceived as the ability to anticipate or even create tomorrows demand. Who would have guessed 15 years ago that teenagers would possibly spend more time sitting at home alone chatting with their friends than doing face-to-face communication?

The innovation cycle in web-based companies is shorter and innovation can be something that we, the users, do not even realize. So can a different programming language reduce the total storage on the server or improve the download speed. The products characteristic derives from the various options the user can choose from. Googles browser for example can get personalized and it is able to connect all services provided by the company. Despite the pure browser options, Google created an interface between the user and the web. Success will soon no longer be measured as the number of clicks or number of members. There might be more value in the total data volume transferred and stored. The information collected by a web site provider can be very powerful. Users unveil their privacy by uploading their lives. Thus, Internet based companies needed new structures to align the business areas they operate in. Unlike a car manufacturer who will remain in the car manufacturing industry. Thus the needs of existent product businesses are totally different. Products for use no matter if real (car) or virtual (bank account) need warranty, contracts and quality inspections.

The traditional business model is founded on making money with the product produced. Traditional industries focus mainly internally on products or processes. The attention is inward rather than outward like in web operating companies. The innovation being directed outside the company can also be used as a marketing tool. Take eBay as an example when it introduced its PayPal service. This financial solution has very little in common with the original trading platform. By launching it as a separate and individual tool, eBay received a lot of attention. The creative part is to imagine what users might want besides the service they came for originally. The focus on core competences is a reason why the traditional industry is rather slow with such products and services. The recently established financial services owned by car manufacturers are just the first examples to align new products with services in the traditional industry. This is learning from the web based companies success.

Sustainability as marketing and sales factor therefore should be seen less ecologically but strategically. Once you sustain in offering an advanced service, continue to be the leader in technology or are recognized as the low price source, you will bind your customers. Over the last 20 – 25 years, the German grocery store Aldi is perceived to be the leader in price. They sustain their lowest cost image and most consumers shop with the positive sureness to get the best price. Do they really? Aldi (North & South) operates in more than 27 countries worldwide and the founding brothers belong to the wealthiest people in Europe. The success is obvious and comes from a highly innovative and flexible organization. The Aldi concept is to sustain the high quality standards of no name brands at low cost. At the same time, the reduced variety of articles improves the turn over and leads to positive cash flow. Continuous and sustainable change, e.g. adding clothes to the product line, enables Aldi to maintain and widen their successful business model.

8.5 Summary

Successful businesses enable their employees to be successful. The workforce will become an irreplaceable asset as modern high tech jobs are related less to manpower but to experience, know-how and emotional commitment. Those companies that retain their brilliant brains and experienced workers have a higher chance for future invention and faster market implementation. It is in the natural interest of any employer to maintain those who have or can gain know-how. The innovative driving force will no longer be the privilege of those in R&D or engineering departments but needs to be spread equally into every single step of the value chain; especially in high labor cost countries everyone is needed as source of creativeness.

As most processes are complex and have hidden issues, mathematical modeling is not always the primary choice as a starting point for change. The change needs to happen within the organization first. The traditional hierarchic constructions are already compressed and many levels have been removed. This was done in the true belief to strengthen the business. But to get more done with less hierarchy requires a different working attitude and altered organizational communication and commitment. Furthermore, it requires a shared vision comprehensible to everybody. The executing force needs to share the vision but also needs to understand what part of the vision is for them. Communication skills, project and people management and the ability to listen and reflect on oneself critically are key necessities for successful and sustainable change. These conditions are true for project managers but even more for the entire organization. Sustainability is obtained if the final users are not only a part of the change right from the beginning, but are the drivers of change and innovation and interact equally and actively.

Modern hierarchic constructions need to turn their hierarchic pyramid upside down and focus on the satisfaction and the interests of every individual. Talent development programs focusing on retaining expertise as well as a strong employer brand to attract talents are key prerequisites for the future. The demographic change forces organizations to reconsider their attitude towards the human aspect. Thus successful businesses prove that non-job related tasks can support the process of creativity and innovation and leads to the above mentioned emotional commitment. Social values, whether it is in flexible working time, parental leave or educational leave, eventually create a different organizational environment. This will turn into the formation and development of a most creative atmosphere, which is the key for innovation and satisfaction. In order to make the innovation last, successful employees in an emancipated hierarchy are necessary. The latter is not by chance but is derived when the top managers accept resistance as a source of employee interaction throughout all hierarchic levels. We need to make use of the diversified strength of every individual but at the same time value the weaknesses of humans. Most modern companies simply cannot afford to refrain from using their human assets intensively.

Focus on, and increased investment in, the human aspect will not only speed up innovation, market implementation and sustainability, it is the most valuable asset of all. Make use of this asset but treat it with respect and do not forget that "there is a difference between listening and waiting to speak!" [61]

References

1. E.H.L. Aarts, P.J.M. Korst, and P.J.M. van Laarhoven. *Pattern Recognition: Theory and Applications*, chapter Simulated Annealing: A Pedestrian Review of the Theory and Some Applications. Springer Verlag, 1987.
2. E.H.L. Aarts, P.J.M. Korst, and P.J.M. van Laarhoven. Quantitative analysis of the statistical cooling algorithm. *Philips J. Res.*, 1987.
3. E.H.L. Aarts and P.J.M. van Laarhoven. A new polynomial time cooling schedule. In *Proc. IEEE Int. Conf. Comp. Aided Design, Santa Clara, November 1985*, page 206 208, 1985.
4. E.H.L. Aarts and P.J.M. van Laarhoven. Statistical cooling: A general approach to combinatorial optimization problems. *Philips J. of Research*, 40:193 226, 1985.
5. Forman S. Acton. *Numerical Methods That Work*. The Mathematical Association of America, 1990.
6. Teresa M. Amabile, Regina Conti, Heather Coon, Jeffrey Lazenby, and Michael Herron. Assessing work environment for creativity. *The Academiy of Management Journal*, 39(5):1154 – 1184, 1996.
7. Patrick D. Bangert. How smooth is space? *Panopticon*, 1:31 – 33, 1997.
8. Patrick D. Bangert. *Algorithmic Problems in the Braid Groups*. PhD thesis, University College London Mathematics Department, 2002.
9. Patrick D. Bangert. Mathematik – was ist das? *Bild der Wissenschaft*, page 10, 2004.
10. Patrick D. Bangert. Raid braid: Fast conjugacy disassembly in braid and other groups. In Quoc Nam Tran, editor, *Proceedings of the 10th International Conference on Applications of Computer Algebra*, ACA, pages 3 – 14, 2004.
11. Patrick D. Bangert. Downhill simplex methods for optimizing simulated annealing are effective. In *Algoritmy 2005*, number 17 in Conference on Scientific Computing, pages 341 – 347, 2005.
12. Patrick D. Bangert. In search of mathematical identity. *MSOR Connections*, 5(4):1 – 3, 2005.
13. Patrick D. Bangert. Optimizing simulated annealing. In *Proceedings of SCI 2005 – The 9th World Multi-Conference on Systemics, Cybernetics and Informatics from 10.-13.07.2005 in Orlando, FL, USA*, volume 3, page 198202, 2005.
14. Patrick D. Bangert. Optimizing simulated annealing. In Nagib Callaos and William Lesso, editors, *9th World Multi-Conference on Systemics, Cybernetics and Informatics*, volume 3, pages 198 – 202, 2005.
15. Patrick D. Bangert. What is mathematics? *Aust. Math. Soc. Gazette*, 32(3):179 – 186, 2005.
16. Patrick D. Bangert. *Jenseits des Verstandes*, chapter Einführung in die buddhistische Meditation, pages 165 – 172. S. Hirzel Verlag, 2007.
17. Patrick D. Bangert. *Jenseits des Verstandes*, chapter Inwieweit kann man mit Logik spirituell sein? Die Sicht eines Mathematikers und Buddhisten, pages 147 – 152. S. Hirzel Verlag, 2007.
18. Patrick D. Bangert. *Kreativität und Innovation*, chapter Kreativität in der deutschen Wirtschaft, pages 79 – 86. S. Hirzel Verlag, 2008.
19. Patrick D. Bangert. Mathematical identity (in greek). *Journal of the Greek Mathematical Society*, 5:22 – 31, 2008.
20. Patrick D. Bangert. *Lectures on Topological Fluid Mechanics*, chapter Braids and Knots, pages 1 – 74. Number 1973 in LNM. Springer Verlag, 2009.
21. Patrick D. Bangert. *Neuroästhetik*, chapter Fraktale Kunst: Eine Einführung, pages 89 – 95. E. A. Seemann, 2009.
22. Patrick D. Bangert. Ausbeuteoptimierung einer silikonproduktion. In *Arbeitskreis Prozessanalytik*, number 6 in Tagung, page 14. DECHEMA, 2010.
23. Patrick D. Bangert. Ausbeuteoptimierung in der silikonproduktion. *Analytic Journal*, page www.analyticjournal.de/ fachreports/ fluessig_analytik/ algorithmica_technol_silikon.html, 2010.

24. Patrick D. Bangert. Increasing energy efficiency using autonomous mathematical modeling. In Victor Risonarta, editor, *Energy Efficiency in Industry*, Technology cooperation and economic benefit of reduction of GHG emissions in Indonesia, pages 80 – 86. Shaker Verlag, 2010.

25. Patrick D. Bangert. Two-day advance prediction of a blade tear on a steam turbine of a coal power plant. In M. Link, editor, *Schwingungsanalyse & Identifikation. VDI-Berichte No. 2093*, pages 175 – 182, 2010.

26. Patrick D. Bangert. Two-day advance prediction of a blade tear on a steam turbine of a coal power plant. In *Instandhaltung 2010*, pages 35 – 44. VDI/VDEh, VDI/VDEh, 2010.

27. Patrick D. Bangert. Prediction of damages on wind power plants. In *Schwingungen von Windenergieanlagen 2011*, number 2123 in VDI Berichte, pages 135 – 144. VDI, 2011.

28. Patrick D. Bangert. Prediction of damages using measurement data. In Bernd Bertsche, editor, *Technische Zuverlässigkeit*, number 2146 in VDI Berichte, pages 305 – 316. VDI, 2011.

29. Patrick D. Bangert. Two-day advance prediction of blade tear on the steam turbine at coal-fired plant. In *54th ISA POWID Symposium*, volume 54 of *ISA*. ISA, 2011.

30. Patrick D. Bangert and Markus Ahorner. Modellierung eines pumpenanlaufs zur lebensdaueroptimierung mit der völlig neuen n-körper methode. In *Produktivitätssteigerung durch Anlagenoptimierung*, number 29 in VDI / VDIEh Forum Instandhaltung, pages 29 – 36. VDI/VDEh, 2008.

31. Patrick D. Bangert, M.A. Berger, and R. Prandi. In search of minimal random braid configurations. *J. Phys. A*, 35:43–59, 2002.

32. Patrick D. Bangert, Mitchel A. Berger, and Rosela Prandi. In search of minimal random braid configurations. *J. Phys. A*, 35:43 – 59, 2002.

33. Patrick D. Bangert, Martin D. Cooper, and S.K. Lamoreaux. Enhancement of superthermal ultracold neutron production by trapping cold neutrons. *Nuc. Instr. Meth. in Phys. Res. A*, 410:264 – 272, 1998.

34. Patrick D. Bangert, Martin D. Cooper, and S.K. Lamoreaux. Uniformity of the magnetic field produced by a cosine magnet with a superconducting shield. *LANL EDM Expt. Tech. Rep.*, 1, 1999.

35. Patrick D. Bangert and Jörg-Andreas Czernitzky. Increase of overall combined-heat-and-power (chp) efficiency via mathematical modeling. In *VGB Fachtagung Dampferzeuger, Industrie- und Heizkraftwerke*, 2010.

36. Patrick D. Bangert and Jörg-Andreas Czernitzky. Efficiency increase of 1% in coal-fired power plants with mathematical optimization. In *54th ISA POWID Symposium*, volume 54 of *ISA*. ISA, 2011.

37. Patrick D. Bangert and Jörg-Andreas Czernitzky. Increase of overall combined-heat-and-power efficiency through mathematical modeling. *VGB PowerTech*, 91(3):55 – 57, 2011.

38. Patrick D. Bangert, Chaodong Tan, Zhang Jie, and Bailiang Liu. Mathematical model using machine learning boosts output offshore china. *World Oil*, 231(11):37 – 40, 2010.

39. Patrick D. Bangert, Chaodong Tan, Bailiang Liu, and Zhang Jie. Maschinelles lernen erhöht ertrag. *China Contact*, 15(6):52 – 54, 2011.

40. D.M. Bates and D.G. Watts. *Nonlinear Regression Analysis and Its Applications*. Wiley, 1988.

41. M.A. Berger. Minimum crossing numbers for three-braids. *J. Phys. A*, 27:6205–6213, 1994.

42. Lutz Beyering. *Individual Marketing*. Verlag Moderne Industrie, 1987.

43. Marco A. D. Bezerra, Leizer Schnitman, M. de A. Baretto Filho, and J.A.M. Felippe de Souza. Pattern recognition for downhold dynamometer card in oil rod pump system using artificial neural networks. *Proceedings of the 11th International Conference on Enterprise Information Systems ICEIS 2009, Milan, Italy*, pages 351 – 355, 2009.

44. C.M. Bishop. *Pattern Recognition and Machine Learning*. Springer Verlag, 2006.

45. Dan Bonachea, Eugene Ingerman, Joshua Levy, and Scott McPeak. An improved adaptive multi-start approach to finding near-optimal solutions to the euclidean tsp. In *Genetic and Evolutionary Computation Conference (GECCO-2000)*, 2000.

46. F. Bonomi and J.-L. Lutton. The asymptotic behaviour of quadratic sum assignment problems: A statistical mechanics approach. *Euro. J. Oper. Res.*, 1984.
47. E. Bonomi and J.-L. Lutton. The *n*-city travelling salesman problem: Statistical mechanics and the metropolis algorithm. *SIAM Rev.*, 26:551 568, 1984.
48. M. Boulle. Khiops: A statistical discretization method of continuous attributes. *Machine Learning*, 55:53 – 69, 2004.
49. Wayne H. Bovey and Andrew Hede. Resistance to organisational change: the role of defence mechanisms. *Journal of Managerial Psychology*, 16(7):534 – 548, 2001.
50. Michael Brusco and Stephanie Stahl. *Branch-and-Bound Applications in Combinatorial Data Analysis*. Springer Verlag, 2005.
51. R.E. Burkard and F. Rendl. A thermodynamically motivated simulation procedure for combinatorial optimization problems. *Euro. J. Oper. Res.*, 17:169 174, 1984.
52. V. Cerny. Thermodynamical approach to the travelling salesman problem: An efficient simulation algorithm. *J. Opt. Theory Appl.*, 45:41 51, 1985.
53. William G. Cochran. *Sampling Techniques*. Wiley, 1977.
54. N.E. Collins, R.W. Eglese, and B.L. Golden. Simulated annealing - an annotated bibliography. *Am. J. Math. Manag. Sci.*, 8:209–307, 1988.
55. Peter Dayan and L. F. Abbott. *Theoretical Neuroscience*. The MIT Press, 2001.
56. John E. Dowling. *Neurons and Networks: An Introduction to Neuroscience*. The Belknap Press of Harvard University Press, 1992.
57. L.A. McGeoch D.S. Johnson, C.R. Aragon and C. Schevon. Optimization by simulated annealing: An experimental evaluation. In *List of Abstracts, Workshop on Statistical Physics in Engineering and Biology, Yorktown Heights, April 1984, revised version.*, 1986.
58. H. DeMan F. Catthoor and J. Vanderwalle. Sailplane: A simulated annealing based cadtool for the analysis of limit-cycle behaviour. In *Proc. IEEE Int. Conf. Comp. Design, Port Chester, Oct. 1985*, page 244 247, 1985.
59. A.L. Sangiovanni-Vincentelli F. Romeo and C. Sechen. Research on simulated annealing at berkely. In *Proc. IEEE Int. Conf. Comp. Design, Port Chester, Nov. 1984*, page 652 657, 1984.
60. U. Fayyad and K. Irani. Multi-interval discretization of continuous-valued attributes for classification learning. In *Proc. of the 13th Int. Joint Conf. on Artificial Intelligence*, pages 1022 – 1027, 1993.
61. Tom Foremsko. Twitter from cocoon.ifs.tuwien.ac.at, 2009.
62. David A. Freedman. *Statistical Models: Theory and Practice*. Cambridge University Press, 2005.
63. S. George. *An Improved Simulated Annealing Algorithm for Solving Spatially Explicit Forest Management Problems*. PhD thesis, Penn. State Uni., 2003.
64. Walton E. Gilbert. An oil-well pump dynagraph. *Production Practice, Shell Oil Company*, pages 94 – 115, 1936.
65. Fred Glover and Manuel Laguna. *Tabu Search*. Kluwer Academic Publishers, 1996.
66. B.L. Golden and C.C. Skiscim. Using simulated annealing to solve routing and location problems. *Nav. Log. Res. Quart.*, 33:261 279, 1986.
67. N. Golyandina, V. Nekrutkin, and A. Zhigljavsky. *Analysis of Time Series Structure: SSA and related techniques*. Chapman and Hall/CRC, 2001.
68. J.W. Greene and K.J. Supowit. Simulated annealing without rejected moves. *IEEE Trans. Comp. Aided Design*, CAD-5:221 – 228, 1986.
69. Martin T. Hagan, Howard B. Demuth, and Mark Beale. *Neural Network Design*. PWS Pub. Co., 1996.
70. B. Hajek. A tutorial survey of theory and application of simulated annealing. In *Proc. 24th Conf. Decision and Control, Ft. Lauderdale, Dec. 1985*, page 755 760, 1985.
71. J.D. Hamilton. *Time Series Analysis*. Princeton University Press, 1994.
72. T. Hastie and P. Simard. Models and metrics for handwritten character recognition. *Statistical Science*, 13(1):54 – 65, 1998.
73. Randy L. Haupt and Sue Ellen Haupt. *Practical Genetic Algorithms*. Wiley-Interscience, 2004.

74. Kenneth M. Heilman, Stephen E. Nadeau, and David O. Beversdorf. Creative innovation: Possible brain mechanisms. *Neurocase*, 9(5):369 – 379, 2003.
75. Klaus Hinkelmann and Oscar Kempthorne. *Design and Analysis of Experiments. I and II.* Wiley, 2008.
76. Douglas R. Hofstadter. *Gödel, Escher, Bach: An Eternal Golden Braid.* Penguin Books, 1979.
77. Torbjörn Idhammar. *Condition Monitoring Standards (4 vols)*. Idcon Inc., 2001-2009.
78. Alexander I. Khinchin and George Gamow. *Mathematical Foundations of Statistical Mechanics*. Dover Publications, 1949.
79. S. Kirkpatrick, C.D. Jr. Gelatt, and M.P. Vecchi. Optimization by simulated annealing. *Science*, 220:671 680, 1983.
80. J. Klos and S. Kobe. *Nonextensive Statistical Mechanics and Its Applications*, chapter Generalized Simulated Annealing Algorithms Using Tsallis Statistics, pages 253 – 258. LNP 560/2001 Springer Verlag, 2001.
81. D.E. Knuth. *Seminumerical Algorithms, 2nd ed., vol. 2 of The Art of Computer Programming*. Addison-Wesley, Reading, MA, USA, 1981.
82. J. Lam and J.-M. Delosme. Logic minimization using simulated annealing. In *Proc. IEEE Int. Conf. Comp. Aided Design, Santa Clara, Nov. 1986*, page 348 351, 1986.
83. Rotislav V. Lapshin. Analytical model for the approximation of hysteresis loop and its application to the scanning tunneling microscope. *Rev. Sci. Instrum.*, 66(9):4718 – 4730, 1995.
84. H.W. Leong and C.L. Liu. Permutation channel routing. In *Proc. IEEE Int. Conf. Comp. Design, Port Chester, Oct. 1985*, page 579 584, 1985.
85. H.W. Leong, D.F. Wong, and C.L. Liu. A simulated annealing channel router. In *Proc. IEEE Int. Conf. Comp. Aided Design, Santa Clara, Nov. 1985*, page 226 229, 1985.
86. S. Lin. Computer solutions for the travelling salesman problem. *Bell Sys. Tech. J.*, 44:2245 2269, 1965.
87. H. R. Lindman. *Analysis of variance in complex experimental designs*. W. H. Freeman & Co. Hillsdale, 1974.
88. David G. Luenberger. *Linear and Nonlinear Programming*. Springer Verlag, 2003.
89. M. Lundy and A. Mees. Convergence of a annealing algorithm. *Math. Prog.*, 34:111 124, 1986.
90. J.-L. Lutton and E. Bonomi. Simulated annealing algorithm for the minimum weighted perfect euclidean matching problem. *R.A.I.R.O. Recherche operationelle*, 20:177 197, 1986.
91. P.S. Mann. *Introductory Statistics*. Wiley, 1995.
92. S. Martin, M. Anderson, I. Salman, V. Lazar, and Patrick David Bangert. Processes contributing to the evolution of two filament channels to global scales. In K. Sankarasubramanian, M. Penn, and A. Pevtsov, editors, *Large Scale Structures and their Role in Solar Activity*, ASP Conference Proceedings Series. Astronomical Society of the Pacific, 2005.
93. W. Mass. Efficient agnostic pac-learning with simple hypotheses. In *Proc. of the 7th ACM Conf. on Computational Learning Theory*, pages 67 – 75, 1994.
94. F. Romeo M.D. Huang and A.L. Sangiovanni-Vincentelli. An efficient general cooling schedule for simulated annealing. In *Proc. IEEE Int. Conf. Comp. Aided Design, Santa Clara, Nov. 1986*, page 381 384, 1986.
95. N. Metropolis, A. Rosenbluth, M. Rosenbluth, A. Teller, and E. Teller. Equation of state calculations by fast computing machines. *J. Chem. Phys.*, 21:1087–1092, 1953.
96. D.C. Montgomery. *Design and Analysis of Experiments*. Wiley, 2000.
97. C.A. Morgenstern and H.D. Shapiro. Chromatic number approximation using simulated annealing. Technical report, CS86-1, Dept. Comp. Sci., Univ. New Mexico., 1986.
98. Leonard K. Nash. *Elements of Chemical Thermodynamics*. Dover Publications, 2005.
99. Taiichi Ohno. *Toyota Production System: Beyond Large-Scale Production*. Productivity Press, 1988.
100. Esin Onbasoglu and Linet Özdamar. Parallel simulated annealing algorithms in global optimization. *Journal of Global Optimization*, 19(1), 2001.
101. R.H.J.M. Otten and L.P.P.P. van Ginneken. Floorplan design using simulated annealing. In *Proc. IEEE Int. Conf. On Comp. Aided Design, Santa Clara, Nov. 1984*, page 96 98, 1984.

102. P. Sibani P. Salamon and R. Frost. *Facts, conjectures, and improvements for simulated annealing*. Society for Industrial and Applied Mathematics, Philadelphia, PA, 2002.

103. Athanasios Papoulis and S. Unnikrishna Pillai. *Probability, Random Variables and Stochastic Processes*. McGraw Hill, 2002.

104. Oliver Penrose. *Foundations of Statistical Mechanics: A Deductive Treatment*. Dover Pub. Inc., Mineola, NY, USA, 2005.

105. George Polya. *How to Solve It*. Princeton University Press, 1957.

106. ProQuest. http://www.umi.com/proquest/.

107. D. Pyle. *Data Preparation for Data Mining*. Morgan Kaufmann, 1999.

108. F. Romeo and A.L. Sangiovanni-Vincentelli. Probabilistic hill climbing algorithms: Properties and applications. In *Proc. 1985 Chapel Hill Conf. VLSI, May 1985*, page 393 417, 1985.

109. B. Rosner. On the detection of many outliers. *Technometrics*, 17:221 – 227, 1975.

110. B. Rosner. Percentage points for a generalized esd many-outlier procedure. *Technometrics*, 25:165 – 172, 1983.

111. Stuart Russell and Peter Norvig. *Artificial Intelligence: A Modern Approach*. Prentice Hall International, 1995.

112. C.D. Jr. Gelatt S. Kirkpatrick and M.P. Vecchi. Optimization by simulated annealing. Technical report, IBM Research Report RC 9355, 1982.

113. S. Sahni S. Nahar and E. Shragowitz. Simulated annealing and combinatorial optimization. In *Proc. 23rd Des. Automation Conf., Las Vegas, June 1986*, page 293 299, 1986.

114. Ken Schwaber. *Agile Project Management with SCRUM*. Microsoft Press, 2004.

115. C. Sechen and A.L. Sangiovanni-Vincentelli. The timber wolf placement and routing package. *IEEE J. Solid State Circuits*, SC-20:510 522, 1985.

116. Amartya K. Sen. *Collective Choice and Social Welfare*. London, 1970.

117. Mike Sharples, David Hogg, Chris Hutchinson, Steve Torrance, and David Young. *Computers and Thought: A Practical Introduction to Artificial Intelligence*. The MIT Press, 1989.

118. J. Shore and Warden S. *The Art of Agile Development*. OReilly Media, Inc., 2008.

119. C.C. Skiscim and B.L. Golden. Optimization by simulated annealing: A preliminary computational study for the tsp. In *NIHE Summer School on Comb. Opt., Dublin.*, 1983.

120. R.F. Stengel. *Optimal Control and Estimation*. Dover Publications, 1994.

121. Chaodong Tan, Patrick D. Bangert, Zhang Jie, and Bailiang Liu. Yield optimization in dagang offshore oilfield (in chinese). *China Petroleum and Chemical Industry*, 237(11):46 – 47, 2010.

122. Lloyd N. Trefethen and David Bau III. *Numerical linear algebra*. Society for Industrial and Applied Mathematics, 1997.

123. P.J.M. van Laarhoven. *Theoretical and Computational Aspects of Simulated Annealing*. Centrum voor Wiskunde en Informatica, 1988.

124. P.J.M. van Laarhoven and E.H.L. Aarts. *Simulated Annealing: Theory and Applications*. D. Reidel, Dordrecht, 1987.

125. R. von Mises. *Probability, Statistics and Truth*. George Allen & Unwin, London, UK, 1957.

126. Dianne Waddell and Amrik S. Sohal. Resistance: a constructive tool for change management. *Management Decision*, 38(8):543 – 548, 1998.

127. W.T. Vellerling W.H. Press, S.A. Teukolsky and B.P. Flannery. *Numerical Recipes in C. 2nd edition*. Cambridge University Press, 1992.

128. S.R. White. Concepts of scale in simulated annealing. In *Proc. IEEE Int. Conf. Comp. Design, Port Chester, Nov. 1984*, page 646 651, 1984.

129. Wikipedia. innovation.

130. www.buildabetterburger.com/burgers/timeline.

131. www.foodreference.com.

132. www.whatscookingamerica.net/History/HamburgerHistory.htm.

Index